Becoming-Into-Others (BIO)

Becoming-Into-Others (BIO)

An Interdisciplinary Approach to Problems Concerning Life

Ana Laura Arellano Zamudio, Jorge Vélez Vega,
David Arellano Gault

ANTHEM PRESS

Anthem Press
An imprint of Wimbledon Publishing Company
www.anthempress.com

This edition first published in UK and USA 2026
by ANTHEM PRESS
75–76 Blackfriars Road, London SE1 8HA, UK
or PO Box 9779, London SW19 7ZG, UK
and
244 Madison Ave #116, New York, NY 10016, USA

British Library Cataloguing-in-Publication Data
A catalogue record for this book is available from the British Library.

Library of Congress Cataloging-in-Publication Data: 2026940297

ISBN-13: 978-1-83999-848-5 (Hbk)
ISBN-10: 1-83999-848-2 (Hbk)

Cover credit: Alberto Andrei Rosu/Shutterstock.com

This title is also available as an eBook.

CONTENTS

ACKNOWLEDGEMENTS

We would like to collectively thank the inspiration and love provided by our beautiful companions in life, those who remain and those who have parted but will never be forgotten: Antar, Gol, Fush, Gala, Lima, Bobby, Kimis, Apa, Félix Senior, Juanis, Pepinillo, Pelucas, Félix Jr., Felucas, Peluquitas, Asdrubal Austreberto and the Thunder Cats herd. Their company and kindness ease the difficulties of living on a wounded world.

Ana would also like to thank her human family, Felisa, David, Laura, Jorge, David Ricardo and Carolina for being the greatest friends and supporters. To Felisa, David and Laura, I would like to extend my absolute gratitude for showing me the importance of love, kindness, patience and care. To Jorge, my thought companion, I thank you for all the passionate discussions and the work invested in creating our more-than-human family, a homely utopia.

Jorge would like to thank his parents, Mireya and Arturo, for their unconditional love across lifetime, as well as Laura and David for all their support in his personal and professional life. Last but not least, I would like to thank Ana for giving me the chance to create a community of life where the seed of BIO grew, allowing us to become-into-others in a more tender, kind and loving human beings with the life that surrounds us.

We would also like to collectively extend our gratefulness to Anthem's editorial team, particularly to Jebaslin Hephzibah, Golda Merline and Phiriya Durairaj for the caring work invested in this book and for believing in the BIO project.

Introduction

BECOMING-INTO-OTHERS (BIO): BETWEEN SCIENCE FICTION AND BIOPOWER

The current environmental panorama unfolds like a triptych presenting the image of catastrophe on the left side; a human society inhabiting another planet on the centre side; and a big question mark on the right side, where many imaginaries could be proposed facing uncertain times. We propose, through this book, to think about the idea of BIO or becoming-into-others, as a response to a present that has been focused on recreating the civilization collapse. Before presenting our proposal, let us take a look at the first two sides of the triptych based on two fictional stories that describe them perfectly. It is worth noting that these stories close the twentieth century and open the twenty-first century, respectively. In consequence, these stories were an effect of a certain apocalyptic imaginary and show a glimpse of future possibilities that human beings may encounter while inhabiting a damaged planet.

Left image: In *Dinosaurs'* last TV series episode, the main character named Earl Jim Sinclair apologizes to his family for starting the end of the world and of dinosaurs' civilization. He explains that his faith in progress and technology, coupled with his lack of respect for nature, caused the disaster. He confesses that although he was in charge of the world, he did not know how to take good care of it. Then, referring to his little baby, he assures that there is little world left for him, his brother and sister to live in. With no other world to move to, all that is left is to wait for nature to recover. However, there is still hope. Hopefully, the dinosaurs' long survival on Earth is an indication that they are unlikely to simply disappear.

This scene was a warning about what human beings were facing at the time the TV show was transmitted. Back to the end of the twentieth century, the 1990s decade was experimenting with the first impulses of sustainable development while also facing other kinds of problems, like poverty and social inequality. At that moment, sustainable development focused on regulating pollution, overpopulation and global warming. While it was not an option to eliminate capitalist production and the use of fossil fuels, the strategy was focused on continuing to produce with sustainable criteria, with the conviction to ensure a future for the next generations. In addition to development and productivity, progress would be disguised under the new cloaks of sustainability. But was this the best response to the environmental catastrophe facing the twentieth century? Thirty-eight years after the publication of *Our Common Future* (the precursor of sustainable development, the Brundtland Report), pollution, population and global warming have been increasing phenomena. Apparently, it was not enough, but at least it was the first wakeup call in the face of the climate catastrophe that was to come.

Evidently, the twentieth century did not face the same scenario as the last episode of *Dinosaurs*, and the twenty-first century is still trying to fight to prevent it. The episode referred to is an invitation, an exhortation, in the first instance, not to place all faith in progress and technology, in the second instance, to respect nature, and, in the third instance, to take care of the world. If the present generation of human beings denies or rejects such exhortation, then preparations must be made to explain to the generations to come why they have so little world to inhabit. The new generations will have to realize that the option will be to learn to live in the catastrophe, if it is not yet possible to travel to other planets. In consequence, new generations will have to live not only on a sordid landscape with scarce resources, but also with few prospects for the future.

The irony of the last episode is palpable. Aimed at a younger public, the meaning of the last episode could go unnoticed. However, it is still appropriate to take it into consideration in order to decipher its meaning. That a dinosaur is explicit in stating that they will not disappear simply because they have existed on the planet for centuries is probably one of the arguments humans most often use to reduce their stress in the face of catastrophe. Humans will disappear, but not all of them. In a sense, they are right, because, according to palaeontology, the meteorite that fell to Earth millions of years ago did not wipe out all dinosaurs. Now, the difference between dinosaurs and modern humans is that the former were unaware of the likelihood that an object from outer space would end their existence, while the latter not only know about the potential extinction they face but are actually causing it. So, the last hope of survival does not reside on the long-standing existence on Earth.

Changing the order of ideas, the last hope resides in the recovery of nature. But how could this happen? Unless a spontaneous event ends human beings' existence overnight, human beings will exist for many more years. However, as scholars in the Anthropocene[1] field claim, human beings are the main force causing environmental catastrophe. So, if they continue to exist, then the catastrophe will continue irrevocably. In consequence, nature cannot recover itself if human beings still inhabit it. Immersed in this situation, human beings should change themselves to change the way of inhabiting Earth. This particular way of inhabiting must consider respect for nature (meaning a better understanding of the role of humans in the maintenance of the ecosystem they form part of) and, above all, different practices of caring. If these considerations come to fruition, then it will be possible for human existence as well as other life forms to have the possibility for surviving and BIO.

Centre image: In *Paradises Lost*, Ursula K. Le Guin develops a story of humans travelling through space trying to find another planet to inhabit. The ship has been on an ongoing voyage for years. In the fifth generation of humans, people do not remember what the planet they come from looks like, feels like, or even what nature is. The residues of nature in the ship are comprised mainly of bacteria, plants and humans, but they are all controlled and set in a limited space to which they owe their survival, the spaceship. Some people on the spaceship even question the fact that nature can even exist in a place like that: "'Nature' was what was substantially previous to control, the raw material for control, or what had escaped from control."[2]

This spaceship, limited in space and perfectly calibrated in resources, is all they have to survive. In order to do so, they keep a balanced population of humans (four thousand) and grow plants only enough to feed the exact number of humans living on the ship. They also have particular ways of dealing with residues, including human bodies, which are sent to the "Life Center" to be recycled by a chemical process in order to be re-used. In this sense, when someone dies, they are still part of their world: "She would still be part of their world, not as a being but as an endless becoming [...]. They were all part of one another. All used and users, all eaters, all eaten."[3]

In this sense, the generations aboard the ship do not know about trash, much less trash dumps and islands. They also know nothing about animals, climate and circadian cycles. "They know what life was inside: light, warmth, breath, companionship. They knew that outside was nothing. The void. Death. Death silent, immediate, absolute."[4] Once the fifth generation of humans in the ship arrive to the destination planet, they find it hard to live on, it's uncomfortable to live with an ever-changing nature around them, animals moving around, plants growing in seasons, dirt sticking to their feet. Many people decide not to get off the ship and continue on an endless voyage, understanding the ship as their world and their final destination.

This story is about a voyage of discovery of a new planet, and, in the meantime, it is a redefinition of what those human lives are, what the planet they left behind was, what nature could be inside a ship and the building up of uncertainty of ever going outside the ship. Even if the story is about reaching another planet, it can also show us how humans adapt to a ship with limited space and resources, in which nature has become an obsolete term. Earth, as much as the spaceship, is a container with limited resources and space for life to thrive on. What if we try to understand Le Guin's story as a metaphor for the Earth we live in, a space-rock/ship? What kinds of practices of recycling, of population planning and resource using could we implement to keep our spaceship working for our joined survival and wellbeing?

Right image: As a triptych, the current environmental panorama could not deny or forget the first two images that constitute the whole picture. After all, it is necessary to unfold a triptych to show all the possibilities that lie within. Take, as an example, the famous Jheronimus Bosch's triptych titled *Garden of Earthly Delights*, which shows Eden (left), the Garden of Earthly Delights (centre) and Hell (right). In this case, the unfolding triptych provides a glance at all the possibilities that lie both in eternity and temporality. Left and right images agree in showing the eternity that human beings can only imagine. The central image displays what occurs in human temporality. So, in a single glance, eternity and temporality merge, as do the ideas of the loss of divine grace and infernal damnation. Fallen in temporality, human beings, not able to return to Eden, only see their excesses of pleasure rewarded with eternal torment. In consequence, human beings inhabiting Hell will have lost two worlds. In this case, the right image represents the final condemnation, as if this were the telos of human existence.

For our triptych, which unfolds in times of the Anthropocene and Global Warming, the right image could be one of this kind, but we propose to think about it in other terms. We could simply swap the positions and put on the right an image that represents the salvation of humanity in paradise. But our thinking is not utopic. We would like to propose another kind of idea. The idea of becoming-into-other that accepts, in the first instance, the catastrophe because it arose from it and, in the second instance, Earth as the place where it is still possible to learn to live in. In this sense, we have the conviction that human beings could inhabit Earth in different ways. To outline the idea of becoming-into-other, let us describe it according to the book we are proposing.

This book, that tries to be a techno-symbiont (an artificial companion of thought), was born through the collaboration of three different perspectives that seek, not only to explain the relationships between life, biopower and science fiction or speculative fiction (sf), but also, in first place to cultivate imagination and creativity towards becoming-with the myriad of beings that compose what we have called the natural world; in second place, to reconnect human beings with nature through technical means; and in third place, how these ideas might be thought of in organizational studies debates.

Although this book is not a scientific treatise, it is supported by different biological discourses that explain the ways in which life changes, biopower operates, and sf authors think about possible futures. The book is guided by a principal idea: to think about ways in which the exercise of biopower transforms life with tendencies to project possibilities of different futures, which in some ways have been visualized by the sf literature.

The book is enriched by critical philosophical traditions that challenge classical humanism, engaging in dialogues principally with the works of Michel Foucault and Donna J. Haraway. Additionally, it also connects with critical animal studies, multi-species studies, diverse feminisms and posthumanist theories. It is inspired by books like Haraway's *Staying with the Trouble*, Sherryl Vint's *Animal Alterity: science fiction and the question of the animal,* Sunaura Taylor's *Beasts of Burden*, Aph Ko and Syl Ko *Aphro-ism: Essays on Pop Culture, Feminism, and Black Veganism from Two Sisters*, Matthew Calarco's *Zoographies*, Catherine Malabu's *Ontology of the Accident*, Michel Hardt and Antonio Negri's *Assembly*, Cary Wolfe's *Before the Law*, Felice Cimatti and Carlo Salzani's *The Biopolitical Animal*, and Wayne Gabardi's *The Next Social Contract: Animals, the Anthropocene, and Biopolitics.*

In the face of what has been termed the global society of extinction (a concept that encapsulates the human ecological footprint's trajectory towards catastrophic futures and its implications in our dynamically challenging present), it is imperative to reassess the relationships we, as living entities, maintain with others. This reevaluation calls for a new ontological paradigm of coexistence, articulated through the concept of BIO. Drawing on themes such as the transformation of life, extinction, symbiosis and BIO, this book aims to explore, analyse and strengthen arguments that advocate for innovative ways of cohabitation with the myriad other inhabitants of Earth.

Amidst the escalating global extinction of species, increasing urbanization, habitat reduction for numerous organisms and pervasive pollution, there is a critical need to continuously engage in thought and action directed towards the regeneration of our collectively imperilled planet. This requires rethinking the potential connections and interconnections that can be fostered to enhance coexistence among all forms of life. As a starting point, we take the following questions as a reference for our reflection processes: How are we to survive the global society of extinction? How can we keep on living on a damaged planet? How do we face the narratives of a very possible catastrophic future and an already changing, difficult present? How can we flourish with the myriad beings still living on this planet? Trying to answer these questions, we discovered other possibilities underlying our BIO proposal.

1. What is BIO?

BIO, usually known as a prefix placed before a whole series of words such as logic, technology, mechanics or mimetics, encompasses and brings together everything that can be recognized as life, as well as the connections and interconnections that living beings establish with the non-living and the non-material. In this way, life could not be known without the multiple connections and interconnections, which at the same time serve as a contrast and reflective counterpoint to life itself. Abstracted from its reality, life seems to be something that exists or simply is. However, as a flow that connects and interconnects diachronically and synchronically, life is something that is always becoming and ceasing to be. It is therefore important to reflect on the prefix bio, understood as life, as a way in which one becomes. Thus, living beings are not something that just exists in a predestined nature, but rather something that changes, as it becomes something else.

If life comes to and ceases to exist, it is due to the different connections and interconnections that arise internally-externally/micro-macroscopically between living things, as well as with their environments that include the non-living and the non-material. Through these connections and interconnections, living beings can change and, in any case, transform themselves. These changes and transformations do not occur in empty space, but within or in the middle of a place (a pond), a container (the stomach of a ruminant) or in a specific area (a coniferous forest). The connections and interconnections that living beings establish can be directed towards something or towards someone, since they do not live outside an environment (like the kingfisher, which relates to trees, lakes or rivers, the fish and crustaceans that it eats, the wind that allows it to move and the water in which it dives).

Hence, rather than living under the subjection of external conditions, living beings have an interest in fostering their connections and interconnections in other ways. In this sense, the concreteness of life, that is, living beings in each particular case, shows in its abstraction that life itself changes and transforms within an environment, by directing itself towards something or someone in which one has a specific interest.

Then, just as living beings are not determined and unchangeable, BIO is not a premade and determined concept, nor is it intended to be. It is an integration of concepts that were to be built upon, bit by bit, word by word, by threading ideas into chapters. BIO started as a seed containing the multifaceted potential of the three words that compose it: becoming, into, and others. The first and last ones are mostly stable yet contested territories. *Becoming*, in a succinct way, means accepting the processual nature of life, opposing views that would understand life and living beings as passive materials with unchanging essences. *Others*, is related to the embracing of diversity and change, where difference does not mean inferior, less important or valuable, less interesting or respectable. Difference is not thought in this book through a dichotomic (the Man and its others) hierarchical scheme of thought that has marked many centuries of human thought.

The word *into*, by contrast, was a much more unstable yet fertile ground. *Into* can mean inside or moving inside or to the middle of a place, container, area, etc.; it can also mean change or changing from one form or condition to another; or it can entail a direction, it can mean towards or moving in the direction of something or someone; and finally, it can mean being interested in or being enthusiastic about something. Into, as a preposition, conveys a relationship that connects becoming and others in many different ways by using its possible meanings: transformation or change, moving towards or inside something/someone and deep interest or engagement. In this sense, BIO can be thought of as three ideas: BIO as changing or transforming into others, as becoming towards, inside and with others and as becoming interested in others.

The meanings of the preposition will appear throughout the book paired with three fundamental themes: transformation of life, extinction of life and symbiosis. Doing so will hopefully lead to reflections that advocate for innovative ways of cohabitating with all inhabitants on Earth. Walking towards a future, we need to rethink the potential connections and interconnections that can be fostered to enhance coexistence among all life forms.

2. How this Book Was Written and Reading Recommendations

Writing this book was a collective experiment where the only rule was to follow the threads laid before us by the BIO seed and its growth axis or themes (transformation, extinction and symbiosis). Each of us threaded differently, with our own experiences and categories as well as our own writing style. In a sense, all of us, the authors of the book, brought with us our own companions of thought, which we applied to the compost pile in which the BIO seed was planted. By this interdisciplinary work, we hope to create a diverse patch of forest that can hopefully engage readers in thinking with BIO in creative ways and invite them to contribute to the discussion and add to the diversity needed for a forest to thrive and grow.

There are two main threads inside this book: the sf chapters and the biopower chapters. The sf chapters of the book, written by Ana Arellano in a feminist methodological and epistemological framework following Haraway's situated knowledge,[5] are fundamentally thought experiments, a work of threading in between sf stories, real-life cases and theories to better comprehend the theme in question. They are not written in a sense of literary critique or an intent to ask something from the story or their author of what they should have written or how they should approach a plot, a narrative or an ecological problem, but with an immense respect and care for the art of writing and particularly the writing of sf literature. There is power in these stories, power in the imagination and creativity that the authors gift us with their writing. Sf is a curious genre of literature, one that has been for a long time secluded from the realist literature genres, thought of only as stories of space cowboys, space wars and stories of conquest. But sf is much more than that, it is a fertile ground for creativity and alternatives. In the words of Octavia Butler:

> Science fiction presents differences, alternatives. Alternative societies, technologies, philosophies, alternative species of intelligent beings. It requires people to use their imaginations, stretch their minds, think. It does not require them to display good judgement or great intelligence or, in fact, lack of prejudice.[6]

To think, to stretch the mind, to peel off the mind is the work of sf writing and reading. And, to follow Virginia Woolf's statement in *Three Guineas*: we must think. We cannot allow ourselves to go with the flow of the world that brought us to the catastrophic ecosystem collapse and extinction we live on today if we are to give hope to the next generations. We must find ways to escape the grip of fear and immobilization, and sf can be a fundamental way to do so. Following Ursula K. Le Guin:

> I think hard times are coming where we will be wanting the voices of writers who can see alternatives to how we live now and can see through our fear-stricken society and their obsessive technologies to other ways of being. And even imagine some real grounds for hope. We will need writers who can remember freedom. [...] We live in capitalism, its power seems inescapable, so did the divine right of kings. Any human power can be resisted and changed by human beings. Resistance and change often begin in art. And very often in our art, the art of words.[7]

In this sense, sf is the work of creativity and freedom, and it can be the place to think and mobilize resistance, by thinking of other ways to live, other ways of relating with each other, to become others. Through this book, sf is thought as Le Guin suggests, as a carrier bag: "A leaf a gourd a shell a net a bag a sling a sack a bottle a pot a box a container. A holder. A recipient."[8] A bag of thoughts, feelings, situations, fictitious people and times that can make us think and expand our ways of living: "A book holds words. Words hold things. They bear meanings. A novel is a medicine bundle, holding things in a particular, powerful relation to one another and to us."[9]

In this book, we call BIO, also as a carrier bag, sf narratives are part of the collection, they are seeds of thought, that help us think through transformation of life, extinction, symbiosis, and BIO. Through them, we do not see technology on any of these processes as the savior or the condemnation; we see the many possibilities technology has, how it has been used and might be used in different situations for different purposes. It is true that many sf narratives have been telling the story of technology as a tragic myth, in which the creation turns on to the creator. But this is not the only way to write and understand sf, there are many sf writers who have thought differently about technology and our symbiosis with it.

For Ursula K. Le Guin, for example,

> If science fiction is the mythology of modern technology, then its myth is tragic. 'Technology', or 'modern-science' (using the words as they are usually used, in an unexamined shorthand standing for the 'hard' sciences and high technology founded upon continuous economic growth), is a heroic undertaking, Herculean, Promethean, conceived as triumph, hence ultimately as tragedy. The fiction embodying this myth will be, and has been, triumphant (Man conquers earth, space, aliens, death, the future, etc.) and tragic (apocalypse, holocaust, then or now).
>
> If however, one avoids the linear, progressive, Time's-(killing)-arrow mode of the Techno-Heroic, and redefines technology and science as primarily cultural carrier bag rather than weapons of domination, one pleasant side effect is that science fiction can be seen as far less rigid, narrow field, not necessarily Promethean or apocalyptic at all, and in fact less a mythological genre than a realistic one.
>
> It is a strange realism, but it is a strange reality.[10]

The stories contained in the carrier bag of BIO are not tragic or triumphant myths of technology. They show the in between, the gray areas, the uncomfortable truths behind the uses of technology that prevail to this day. They also show alternatives in the way Octavia Butler suggests, they are ways of creating new ways to live, that tell us who we are, who we hope to be and fear to be. They tell us "[w]hat is in fact going on, what people actually do and feel, how people relate to everything else in this vast sack, this belly of the universe, this womb of things to be and tomb of things that were, this unending story."[11]

Sf, in this sense, is not separate from reality, even if the ways the stories unfold are strange and future focused. They are seeds of our times, and if they grow within us, they can help us, as Haraway suggests, to "craft conditions for finite flourishing on terra, on earth."[12] In this way, sf seeds in the chapters of this book, help us illuminate paths to BIO.

Through the biopower chapters of this book, written by Jorge Vélez, BIO is irrevocably related with biopower or the so-called exercise of power over life. Since Michel Foucault analysed its historical emergence along seventeenth, eighteenth and nineteenth centuries, the relation between life and power has been at stake. Focused on investing and transforming life, biopower has become an important issue in recent times, because it has transcended the very human realm for exerting in different fields, such as animal farming, laboratory experimentation, conservation, rewilding and bio art. More than ever, biopower has taken life itself broadly.

In certain times, from an anthropocentric perspective,[13] it was thought that biopower exerts over human life. But now that prefix "bio-" has opened as a flower, it relates to other life forms, such as other animals, plants, bacteria, even viruses, as well as ecosystems and full environments. The era of biopower designated by Foucault has started with human life control and, eventually, extended to the control of other lives or perhaps in the era of biopower life itself has been controlled since the beginning but the analysis due to their anthropocentric perspective was focused on human life.[14] Despite both directions of analysis, it is settled that biopower has produced different effects on Western civilization and beyond, whether the exercise of power over life learned how to regulate life itself in order to control human life or by controlling human life to regulate other life forms. In this case, what it is necessary to know is that relations with life itself and living beings have always existed, but power relations started to be exerted, Foucault claims, in different ways with the entry of life in the order of knowledge and power in the intersection of eighteenth and nineteenth centuries.

So, the general idea of what life is can find another discursive regime in the epistemic change from Natural History to Biology. Besides knowing what life is, this science can explain life itself, different life forms, within those identified as animals with which humans can be related to. Within this biological frame, human life started to be considered not only just as a living entity, but as a living species among others. Following that idea, human beings understood as living entities have biological mechanisms susceptible to be managed, controlled and regulated. In this sense, just like every animal, human beings are born, get sick, reproduce and eventually die. The novelty of this knowledge is the power effects displayed in all fields where the relationship between life and living entities is of most importance.

In the era of biopower, it is worth highlighting that every living entity *hides* its biological mechanisms that simply work, but once they are discovered and studied, it is possible to control them. Then, that animals have a biological existence is a natural fact that can be administered in certain populations. For example, births can be encouraged, limited, or disallowed; endemic diseases can be treated according to the seasonal periods in which they manifest most acutely; reproduction can be controlled through state planning; and death, as a natural fact, enters in different statistical record because the introduction of a new health campaign could have extended life expectancy. These examples show that life, in this case animal life, has entered a new regime by which it is calculated, managed, administered and controlled by political domains. In consequence, biopower does not constrain its exercise to human life.

Now, the effects of exercise of power over life can be recorded across time. There are three temporal dimensions that can be analysed focusing on how different technologies of power work. Foucault identified three main domains by which the exercise of biopower can be analysed: biohistory, biopolitics and biopoetics. First, biohistory studied chronologically the events in which life and history interfere with one another; second, biopolitics transforms life according to the calculations derived from knowledge–power relationship; and third, biopoetics is related, in first instance, to the possibility of design moral-aesthetic individuality against the biopolitical control of populations and, in second instance, to the possibility of design the life and history interference by which the processes of life could become a historical movement. Each domain is related to a temporal dimension, but they are intertwined. Analysing these three domains' human beings can know what the exercise of biopower has done to their lives and, in turn, what they could stop being in order to become something different.

In a radical sense, if biopower is not exerted only over human life, then, knowing what biopower has done with life, it is possible to change and transform the relationship with life itself. More than ever, in times of Anthropocene, Technocene,[15] Capitalocene,[16] or Plantationocene[17] and Global Warming or Global Boiling, a new interference between life and human history must be designed to make human beings and life itself survive. Following this chain of arguments, BIO or becoming-into-others, based on certain historical analyses of the exercise of power, is charged with all the biopoetic power that inspires a change in contemporary lifestyles whose effects allow both human beings and life itself to become different. In general, this proposal, linking biohistory, biopolitics and biopoetics, aims to allow life to come into being in a different world, thinking of new ways of interference and connections between life and history.

Lastly, David Arellano, brings forth the knowledge of Organizational Theories in link with the BIO proposal. The importance of this framework is to find ways to create a more practical BIO in the way humans organize and the way we understand organizations to be and to develop. In this sense, the chapter contains a series of discussions and proposes suggestive bibliography to work with. The chapter works as a proposal of how BIO could be inserted in Organizational Theories as a fundamental concept of becoming-into-other ways of organizations.

For our readers, we believe that given the depth and specialization of the discussions presented within this book, it is primarily targeted towards an academic audience, encompassing both scholars and students at the undergraduate and graduate levels. We trust it would be of particular interest for those engaged in critical studies, philosophy, cultural studies, gender studies, literature, animal studies, environmental studies and critical public policy studies. The interdisciplinary nature of the book makes it a valuable resource for those interested in exploring complex themes across various domains of inquiry, fostering a nuanced understanding of the topics discussed. We hope this book invites readers to reflect not only theoretically about the subjects at hand but also inspires them towards action and practices that may benefit them, the ones who surround them and the environment they inhabit.

Even though we wrote each chapter on an individual manner, our relationship as authors of this book created a rich environment for discussion and thought throughout the whole process, which is why the threads of the book align and complement each other. We believe it was important to have individual freedom to write and argue in our own terms, but the book in its entirety is worked by all of us in a collective and profound way.

This arrangement also allows readers to select how they wish to read the book. Even if it is being presented as a sequence, the book can be read in different orders. Some who might be more engaged with the biopower discussions may want to read those first and then complement with the other chapters; this would not be a problem since the chapters are self contained and may be read independently. In this sense, we would like to provide readers small chapter summaries to get acquainted and excited about the book's ideas to follow.

3. Chapter Summaries

This book is composed of five parts related to four guiding themes: transformation, extinction, symbiosis and BIO. Each part is divided into two chapters, the first chapter on each part is focused on thinking with sf narratives, the second chapter delves on theoretical and empirical cases that show the exercise of power over life.

Part I. Transformation of life

From the realms of science fiction to the empirical rigors of scientific experimentation (spanning laboratories to animal farms) the past century has underscored that life is not merely an ahistorical essence inherent to all living organisms but is instead a historical object susceptible to transformation through what Foucault has termed biopower, or the exercise of power over life. Although the manifestations of this power may outpace the gradual alterations wrought by evolutionary mutations, distinguishing between artificial and natural selection, the subtlety of these changes can render them inconspicuous to the public. Nonetheless, this exercise of biopower not only transforms life but also inaugurates a novel dimension wherein living beings may engage with other forms of life, biotechnologies and materials. This interaction provides the potential for entities to become-into-others or, perhaps, to come into being anew, exemplified whimsically by the creation of a fluorescent bunny. Such instances reveal the profound implications of biotechnological advancements in redefining the boundaries and possibilities of life itself.

In tone with this discussion, Chapter 1, *Transformation of Life: Human-Animal Conundrums of Uplifting Humans, Bodies, Capacities, Power and Otherness,* focuses on three sf stories, all of which explore a specific uplift narrative in which the faulty animal to "enhance" is none other than the human being. These narratives, of course, are embedded in power dynamics and ideas of hierarchy, they show the anxieties of pureness from conservative perspectives embodied by the Ina species (vampire-like symbionts) of *Fledging*, by Octavia Butler. On the contrary, and also by Octavia Butler, the narrative *Wildseed* explores the communities of humans with sets of special powers created by selective breeding and the obsession with enhancement by a man named Doro. Finally, the narrative *Oryx and Crake*, by Margareth Atwood, shows a genetic engineering enhancement focused on survival of a new humanoid species called the Crakers. Through these narratives, the into preposition is clearly linked with change by the idea of transformation of life and towards. Enhancement narratives exist with a towards that guides the processes that these humans live through to get to a goal. Through these narratives we can see different towards unfolding in the same idea of transforming life, specifically humans.

Starting from the possibility of creating human beings from new biotechnologies and different scientific advances that have made it possible to control cellular processes (transform any cell into a stem cell to, eventually, turn into a gamete), Chapter 2, titled *The Transformation of Life in the Age of Biopower*, is focused on exploring the way in which life takes place and the way in which life becomes into other. Despite the lack of a final and a definitive definition of life, it is stated that life changes and transforms (from natural to unnatural) thanks to sciences and biotechnologies. As such, this assumption is another previous or preliminary condition that allows any subject not only to think about the change and transformation of life, but to make it real in any well-controlled environment. However, that processes of life can be changed and transformed not only for scientific feats, but by an exercise of power over life. Within this theoretical reflection, this chapter is meant to examine the relation between life, change and the transformation of life in biopower age setting a question concerning life related with the improbable definition of it; explaining metaphysical, materialistic and biological views about the change and transformation of life; and understanding the contemporary exercise of biopower responsible for not only transforming life but also producing a form of life capable of BIO.

Part II. Extinction of life

In the context of the Anthropocene (a geological epoch marked by human influence) it becomes evident that the existential threat of extinction facing all life forms is intricately linked to the cumulative human footprint. This footprint comprises a range of detrimental impacts, including the accumulation of atmospheric CO_2, the pervasive remnants of technical fossils (a euphemistic term for trash), widespread deforestation, habitat desertification and the proliferation of persistent new materials such as microplastics. Within this framework of human ecology, there is an urgent need to reconceptualize our planetary cohabitation. This entails devising innovative ways of living that support and enhance the ability of various entities to become-into-others and come into being, thereby averting a potential mass extinction of multiple species, not solely our own.

Hence, Chapter 3, *Extinction or Twisting Towards Renewal of Life on Earth*, explores Sunaura Taylor's disabled ecologies, while framing it through the lens of the processes of extinction we face today. Through this framework, it is fundamental to show the importance of holding accountability in the reasons why ecosystems, populations, communities and species go extinct or are thrown towards extinction. Also, it proposes ways of mitigating, supporting and healing those places and beings; by centering the deep importance of the loss of a partner organism, loss of a home or spaces to live in.

Through *Dawn*'s narrative, by Octavia E. Butler, the chapter delves into conservation practices focused on breeding and hybridization. Lilith, a woman taken by an alien species named Oankali, is one of the humans rescued from the destroyed Earth to be saved from extinction. This story, compared to conservation practices actually on Earth today, shows how much we humans act like the Oankali, in how we use individuals of species to breed them back from endangerment, without a question for what those individuals may be suffering in the process. The second story, *Barn 8*, by Deb Olin Unferth, takes us inside Barn Universe where chickens are constantly modified, through several techniques, especially reproductive technologies, to be more productive. These animals are not endangered whatsoever, if anything they are overbred. When they are freed, they are able to survive a damaged and rough environment because of their resilience to the stress they are accustomed to. Through these narratives, important questions arise about extinction and the renewal of life on Earth. How, when and why to intervene (mitigate, support, heal)? Why do some animals deserve to be saved and how much do they lose when they are subjected to conservation practices?

Questioning certain human practices that throw species into extinction, Chapter 4, titled *Extinction: from Biopower to Biopoetics of Life*, examines the productive dimension of biopower that makes live, lets die and throws into death and extinction. Now that humanity is facing a series of species extinction caused by anthropogenic activities linked to global warming, can the situation be changed? Or is humanity condemned to see how different species become extinct? Is it possible to change the biohistorical record of extinction? How to create another perspective that resists biopolitics and transforms biohistorical interference between life and history? In front of biopolitics and biohistory, is it possible to put a form of biopoetics, understood as the way by which interference between life and history would happen in the future?

To try and answer these questions, the chapter shows that biopower makes live, lets die, throws to death and extinction; the importance, in biohistorical terms, of lists and records of extinct species, as well as the problems they face that render them inoperative to some extent, to demonstrate the ways in which the movements of life and the processes of history interfere with each other; and, working with the interference between life and history, the gallery of biopoetic perspectives from different theoretical fields to propose a way in which the processes of life could become a historical movement to avoid extinction.

Part III. Symbiosis

Symbiosis, as a scientific term, delineates the relationships between organisms of different species, it is also a term that has provided explanations for two significant evolutionary theories. The first is Lynn Margulis' theory of endosymbiosis, which posits the origin of eukaryotic cells as a quintessential example of becoming-into-others. The second, symbiogenesis, emphasizes the enduring impact of symbiotic relationships on the evolutionary trajectory and, consequently, the persistence of life. These theories collectively enrich our understanding of evolution, illustrating that the essence of evolutionary success lies in the capacity for BIO.

Chapter 5, *Symbiosis and the Sym-imaginaries of Becoming-with*, focuses on an exploration of symbiosis, symbiogenesis and sympoiesis frameworks, as a critique of individual and progress ontologies and the anthropocentrism embedded in scientific thought; and as a proposal towards understanding we are all made by and with others in a complex world of relationships. In that sense, sym-imaginaries would propose less parasitic and anthropocentric ways to relate with others; as well as a constant exercising of our response-abilities, towards sympoietic ways of living and dying.

Through *Indra's Web*, by author Vandana Singh, a story about sympoietic terraforming and response-ability, the main character, Mahua, embodies an attentive and caring way of relating that can be thought through sym-imaginaries. These sym-imaginaries closely relate to the into proposition sense of interest and attachment. It is an active, productive and creative proposal: taking interest in others in a proactive way, where attachments, in Isabelle Stengers' sense (feelings and thoughts that compel us to act), do not paralyze us but exert us to keep on an interest and being active about it in maybe small, seemingly inconsequential practices on the global scale; but consequential to our personal and immediate relationships becoming's.

Chapter 6, titled *Symbiosis: from Biopolitics to Symbiopolitics and Beyond*, technosymbiosis examines the intimate relationship between technology and symbiosis. Technosymbiosis is the way by which human beings could become into other in relation to their symbionts. Given the fact that symbiosis is a process of life that has become relevant in the las two decades, it is important to analyse the biohistorical effects produced in the age of biopower. In this sense, the main objective of this chapter is to analyse the ways in which biopower transform life when it considers the biological process of symbiosis. This chapter describes the way in which human beings are understood as symbionts in the age of biopower; the biohistorical conditions under which humans have made symbiosis with other species; technology as a human symbiont to explain other forms of coevolution; the biopolitical effects of discourses on symbiosis in the case of symbiotic engineering applied in the animal industry; and how the scientific discourse of symbiosis can configure a form of resistance to the power over life that instrumentalizes and exploits life itself.

Part IV. Becoming-into-others

In light of the catastrophic impacts' humanity has exerted on our shared planet, Donna J. Haraway's work emerges as a crucial intellectual resource for rethinking and reconstructing our relationships across species and with our environment. Becoming-with, a concept Haraway has worked with encourages an embodied presence in the mud, the thick present of learning from others around us. Hailing towards becoming-with others in order, not only survive, but to flourish through the cultivation of the arts of living and dying on a planet marked by ecological disruption. Becoming-with, in this sense, inspires a becoming-into-other's framework that tends towards more mutualistic practices and accountabilities with the world and others around us.

Chapter 7, *Becoming-Into-Others, Becoming Earthseed,* delves into two attachments or situated perspectives that are important for the BIO framework in development. One of them focuses on Earthseed's philosophy in the novels *Parable of the Sower* and *Parable of the Talents,* by Octavia Butler, in which Lauren Oya Olamina, the main character, developed her way of thinking about the world she lives in and about the changing nature of life. The chapter involves a threading of some of her quotes with the idea of BIO. The second attachment is a narrative from Ana's personal experience in which many animals come to play, as well as the changing environment of a fragment of Mexico City. Of course, as a personal story is not intended to show a recipe for success, it is not even a story with a clear-cut conclusion of how to treat animals in liminal spaces. The purpose of this personal story is to show first how BIO can be applied into a small and personal perspective; second how this framework may guide us through difficult moments and hard decisions and keep us asking questions about the world around us, to keep seeking new ways of responding to the living beings we come into contact with, in our everyday lives.

Chapter 8, titled *Becoming-Into-Others: a Symbiopoetic Approach to Reconnection with Nature,* aims to design a symbiopoetic BIO, that is focused on transforming the interference between life processes and historical movements according with symbiotic power of life. This approach must consider both biological and technical symbiosis in order to think a new way to reconnect human beings with other organisms, environments and planet Earth. This chapter relates the problem with plastics to a sf story to describe the monstrous effects of technological products and the technosymbiotic human condition; examines the historical conditions by which humans have disconnected from nature and the technical reconnection possibility; explains four cases in which the connection, disconnection and reconnection of nature and humans are shown; and determines the conditions under which humans can imitate the power of life, considering the life-giving power of trees, to reconnect with nature.

Part V. Organizing-into-others

The framework of BIO calls for innovative paths in public policy, new ways of organizing and a reimagined approach to international cooperation. The book underscores the urgent need for new governance frameworks, as global crises such as climate change, social inequity and international conflicts intensify. It challenges traditional, human-centric and nation-centric approaches in public policy by promoting more holistic, ecologically informed perspectives. These perspectives acknowledge the interconnectedness of human, non-human and environmental stakeholders, arguing for policies that are not merely reactive and remedial but fundamentally preventative and cooperative. Such policies would embrace principles of symbiosis and mutualism (mutually beneficial associations) across species and ecosystems.

Chapter 9, *Organizing-Into-Others: BIO, Anarchy and the Ethics of Collective Becoming,* addresses the existing organizational paradigms in both governmental and non-governmental sectors, which often perpetuate silos and inefficiencies, are critiqued for their inability to effectively respond to these complex societal challenges. BIO proposes that organizations adopt more fluid and adaptable forms, which can better meet the dynamic nature of global challenges. By integrating principles from biological ecosystems, such as diversity, adaptability and resilience, organizations can evolve towards more sustainable and responsive operational modalities.

A significant challenge outlined in the book is the transition towards more participatory and inclusive forms of governance that actively engage diverse community voices, including those traditionally marginalized. The goal is to foster organizations that are not only open and democratic but also effective and transparent. This comprehensive rethinking of organization and organizing is essential for addressing the multifaceted issues our world faces today.

Notes

1 The concept of Anthropocene was proposed by Paul J. Crutzen and Eugene Stoermer in 2000 to designate a historical change in which human beings are considered a geological force capable of changing the face of the Earth significantly. Global Warming, as an atmospheric phenomenon, is the paradigmatic case of this new geological era that is leaving behind the Holocene. Birgit Schneider has made visible the change through the comparison of the famous image of "The Blue Marvel" with the "burning world images. Passing from color blue to red the images show the evidence of the human footprint in climatic terms. See: Brigit Schenider, "Burning worlds of cartography: a critical approach to climate cosmograms of the Anthropocene", *Geo: Geography and Environment 3*, no. 2 (2016), https://doi.org/10.1002/geo2.27.

2 Ursula K. Le Guin, *The Birthday of the World and Other Stories* (New York: Perennial HarperCollins Publishers, 2002), 276.
3 Ibid., 262.
4 Ibid., 289.
5 This is why the sf chapters are written in a personal style.
6 Octavia E. Butler, *Octavia E. Butler: The Last Interview and Other Conversations* (Melville: Melville House Printing, 2023), 17.
7 Ursula K. Le Guin, "Speech in regards of accepting the National Book Foundation's Medal for Distinguished Contribution to American Letters", New York City, November 19, 2014, American Masters PBS https://www.youtube.com/watch?v=s2v7RDyo7os
8 Ursula K. Le Guin, *The Carrier Bag Theory of Fiction* (UK: T. J. Books Limited, 2019), 29.
9 Ibid., 34.
10 Ibid., 36.
11 Ibid., 37.
12 Donna Haraway, *Staying with the trouble: Making kin in the Chthulucene* (Durham: Duke University Press, 2016), 10.
13 An anthropocentric perspective is a worldview in which humans are the center of interest.
14 Although Foucault's biopower framework has been regarded as anthropocentric, throughout this book, especially on the themes related to biopower, Foucault's conceptual toolbox is employed to analyze different problems regarding life in general beyond the anthropocentric perspective. To know more about debates around this topic, see Chapter 8, subtitled The Technospheric Hyperconnection. Furthermore, we start from the analysis of the effects produced by the era of biopower rather than the epoch called Anthropocene because, on the one hand, if power produces knowledge, then biopower, understood as the administration of life started in 18th century, has set the conditions for scholarly work to formalize the knowledge of the Anthropocene at the beginning of the 20th century; on the other hand, if the Anthropocene were to come to an end, which would be the dream of certain posthumanist intellectual personalities, the exercise of power over life would continue by other means.
15 Contrary to the Anthropocene, which claims that human beings are the cause of the environmental crisis, Technocene focuses its attention on the effects produced by technology since the Industrial Revolution, when technology started to impact on planetary scale.
16 Capitalocene proposes an era in which climate change and other environmental crisis are not only created by human being's activities, but also or more importantly because of the means of capitalistic production.
17 Plantationocene suggests that the era of environmental and social crisis we live in has been created by the colonial and imperialistic systems that include plantations, slavery, forced labor and objectification of nature.

Part I

TRANSFORMATION OF LIFE OR THE CHANGES IN LIFE'S DESIGN

Chapter 1

TRANSFORMATION OF LIFE: HUMAN–ANIMAL CONUNDRUMS OF UPLIFTING HUMANS, BODIES, CAPACITIES, POWER AND OTHERNESS

All that you touch
You Change.
All that you Change
Changes you.
The only lasting truth
Is Change.

—Octavia E. Butler,
Parable of the Sower

Following Earthseed ideas in Octavia Butler's *Parable of the Sower*: Life is change. From subtle changes in biochemical processes and bodily reconfigurations, to generational evolutionary change; life as we know it is bound to transform. Animals, protozoans, fungi, plants, bacteria, algae and everything in between categories of living beings are constantly changing themselves and each other in profound ways. Individuals change with their environment, as well as change their environment. Life is intertwined in many perceptible and imperceptible ways. From organelles, cells, tissues, organs, systems, organisms, population, communities and so on; all living beings partake in small- and large-scale processes that are in play at different speeds and depths.

There is no way to escape the processes of transformation we collectively partake on. In the midst of the rise of the holobiont, understood as a collective of beings, symbiogenesis theory has gained attention. Initially described by plant physiologist Boris Mikhaylovich Kozo-Polyansky in 1924, and later studied by Lynn Margulis, symbiogenesis is a theory of how deeply life on Earth is connected. Kozo-Polyansky pointed to many similarities of organelles, particularly chloroplasts and mitochondria, to ancient bacteria configurations. From these findings, as well as her own, Margulis proposed the theory of endo-genesis, that all eukaryotic (nucleated) cells come from the deeply embedded symbiosis of two bacteria, one who ingested the other, but did not digest it.

These findings made Margulis sustain that the transformations created by the interaction of at least two different species, on a large scale and in the long run, are fundamental processes in evolution. Sometimes even more so than mutation[1] and other processes of speciation[2] considered by evolutionary theorists of her time. This means, as living beings relate to each other in profound ways (even to the point of ingestion but not digestion), they transform each other and, seen from a distance, this is what has created the huge diversity of life we experience today.

As apes and holobionts with our myriads of gut bacteria and part of the community of living beings on this planet, *Homo sapiens* also transform and are transformed. But the large scale of transformation our species have created in the world has been tremendous enough to be called a geologic era (Anthropocene). With a very large population spread throughout the world, written stories that help us understand how we became what we are now, and having created several techniques like science, language, engineering, architecture, education and so on; humans do not only transform others by means of symbiosis, but they also learned to direct change in many other ways.

With all the years of accumulated knowledge and practice regarding transforming life, from artificial selection to genetically targeted manipulation, it is now very common to find easy manuals and instructive videos on how to genetically manipulate animals, plants, fungi, bacteria and any living being we can imagine. There are also protocols and catalogues developed by the animal and agricultural industries. And material evidence in the form of dogs, cats, horses, rats, mice, camels, bunnies, chimpanzees and many other animals that have been transformed by experimental and now standardized protocols of genetic manipulation. Some might say we have taken it too far, others, not far enough.

Because transformation of life has been, in many ways, used as an intensely productive process to generate wealth, many industries have focused their efforts on the reproduction and modification of living beings, either to produce more meat (like the Belgium blue cattle), more eggs or more milk. To create bigger crops, more nutritious meals and many other transgenic benefits (and not so beneficial characteristics) they have encountered along the way. We live in a world full of beings created by a system focused on directing the transformation of life.

Beings inside industries are now even thought of as another kind, different from their animal or plant congeners in nature. Liza Bauer, in her book

Livestock and Literature: Reimagining Postanimal Companion Species, proposed to use the term "postanimal" to explain how animals have been transformed so radically that they cannot still be compared to their natural counterparts, or the animals they used to be before the technological intervention of industry processes. Bauer centers specifically on animals residing in science fiction or speculative fiction (sf) literature to propose the term postanimal, but it can also be brought into reality, in which many transformations have been made on livestock animals or laboratory animals.

In this sense, her analysis of sf narratives that explore what transformations the particular lives of animals in the industry might shine a light to what transformations are going on right now and what they can become. Whether change is brought by the possibilities of exploring space, adjusting to new environments, transformation from organic to machine or directed transformation made by humans or other intelligent species; sf narratives have not been shy of the themes of transformation of life.

These narratives have evolved over time with and sometimes challenging the scientific knowledge of the era, they have become more precise and intrusive in different manners. From *Frankenstein; or, The Modern Prometheus* and *The Island of Doctor Moreau*, where transformation was led by a scalpel; and more recently narratives such as *Jurassic Park* and *MaddAdam*'s trilogy, where transformation is led by genetic manipulation of beings. Particularly, uplift narratives have been very prolific and productive thought experiments.

Uplift involves stories in which the author reflects theoretically on the directed transformation of animals through different technological means, with an idea of enhancement or augmentation of their capacities. This process has a clear *purpose*: making animals more like humans or having human capacities. And even though these narratives have been critiqued for the anthropocentrism underneath their suppositions; to David Brin, the foundational author of the *Uplift Universes* series (1980–1998), uplift is a process humans have not only bestowed on other animals, but also on themselves. In this sense, uplifting processes may include humans, but if uplift's intention is the transformation of animals towards humanness, then, what is the direction for human uplift?

Is there a spectrum of humanness? Are the uplifted humans more human than the ones who are not? Who defines what being more human or a better human means, why and at what cost? There are no simple answers to these questions. Becoming more human, more capable or better, is not a linear and absolute model, it depends on what kind of world you are in and in what kinds of relationships.

As I have tried to entail earlier, transforming life is not a human ordeal only, Margulis and many others have pointed out how bacteria did it first.

We can transform life, because we have been transformed first by tiny microorganisms that shape us to this very moment. But we, as the collective humans, have refined the technique guided by the scientific method[3] and biotechnology advancements. Also, we have added our own views, practices and structures of power, understood particularly as dominion and control of others. When power is tied to transforming life, many issues arise.

The objective of this chapter is to delve into four narratives to better understand these issues as well as the issue of the hierarchy and power that lures close distance. The method I will follow is the string figures games Haraway proposes in her book *Staying with the Trouble*, as they allow and even rely on making things muddy. In this case, making human and humanness muddy enough that, once dried, we can see through the cracks. Maybe as we understand ourselves through these cracks we are better able to become into others in different, more positive and mutualistic ways.

From the standpoint of powerful symbionts of humans in *Fledging*, to Doro's selective breeding in *Wild Seed*; to the genetic manipulation of beings to create a less flawed humanoid in *Oryx and Crake* novel; there is not only a difference in the depth and methods of the transformation of bodies, but also, a parallel evolving of what being human and animal means. They also show how, even as we idolize and exaggerate the importance of our, allegedly, only human capacities, there is a latent idea that humans are faulty organisms who can be improved, sometimes meaning, more powerful, with better senses and mobility, or better adapted to a new world.

1. On the Emergence of a New Species and the Threatening of Taxonomy, Hierarchy and Purity in *Fledging*

Fledging, by Octavia E Butler, as the name suggests, is a story of the emergence of something entirely new, the story of a guided speciation, or maybe the story of two species becoming into something else. The story starts with Shori, a human-like being from a different species called Ina. The Ina have developed symbiotic relationships with humans for centuries, but they are not able to interbreed. Ina feed on human blood, but unlike vampires, they do not often kill humans, in fact, their symbiosis is a mutualistic one, while feeding they provide pleasure, anti-ageing and healing agents through their bloodstream. Humans living with Ina know and accept this trade, as they have for many years.

Just as vampires do, Ina are nocturnal because they lack melanin, hence they have issues with sunlight. Shori comes from a genetic engineering experiment crossing human and Ina DNA. This experiment is not well seen by the most conservative Ina relatives: they believe Ina should be a pure species and so, they decide to kill Shori and her family. She survives but her memory is impaired, she doesn't remember who she is. Throughout the novel, Shori gets to learn how and why her family was slaughtered. With help of other Ina she has found along the way, and with her human symbionts, she prepares to stand trial to get justice for her family.

Ina have lived alongside humans for so long, they are a very similar species, morphologically they are almost human, very pale ones though. They also learned with humans to value reason and use the same human epistemological scales of hierarchy, with a major difference: they believe to be the pinnacle of that scale. At least some of them do, other Ina understand and value the symbiotic nature of the relationship. They recognize the key differences between the species, but doing so does not impair them from loving, caring and even mourning the human companions they live with.

Even though humans and Ina need each other to survive, conservative families show how much superior they believe to be in comparison to humans, and how much Ina-ness should be protected. For once, it is not humans who are superior, they are thought of as less than by a partner they thought of as symbiotic. Why are Ina superior? Well, it is not completely clear, and I believe this is deliberate. To understand it, we might have to appeal to how humans construct hierarchical schemes in the world we live in. Considering that Ina and humans share so much history, it is possible to understand Inas' construction of hierarchy as an actualization of the schemes already put in place.

In the world we know, humans have constructed an idea of human superiority over other animals, and many other dichotomies of hierarchy like culture over nature, white people over black people, man over woman, cis over trans, and so on. One important issue that can be learned from the currents of feminism that scattered during the third wave feminism is that hierarchies are constantly cited and produced. Specifically, I would like to concentrate on understanding how hierarchies are built upon alleged capacities, an idea common in animal studies and in critical disability studies, that can easily be translated to other fields. Let us unravel this idea.

Being human is a hierarchical construction based on capacities. In her book *Simians, Cyborgs and Women*, Haraway conceptualizes what she calls human engineering, that explains how the human category is constructed discursively and materially through different science discourses, particularly biology. This scientific construction of humans, based on capacities that all are supposed to have and can be measured, is an excellent way of excluding everyone who does not have these capacities. Which is especially problematic since, in the world we have inherited, being human ties people to laws and rights that are supposed to protect them.

The second term, human exceptionalism, tends to this exclusion. For Haraway, it means that at the same time the concept of human is constructed under the idea of having certain capacities that can be measured and proved, it is also shaping our understanding of other living beings who are not considered human by the standards of science, or other regimes of truth at the time of the evaluation. Haraway understands human exceptionalism as a rigged game, in which only humans have a set of special characteristics that no other living being has or can obtain.

Of course, interspecies and multispecies, as well as human animal studies, critical animal studies and even science welfare studies and ethology (and many others), have made important contributions to understanding this arranged game of human exceptionalism. For many of these studies, the point is not that there are absolutely no specific abilities that humans have, in comparison to other animals; the point is that all species have specific abilities, special capacities, diverse worlds and possibilities. But that, human exceptionalism narratives are used to justify human hierarchy over other beings.

In this sense, *Fledging*'s narrative experiment of thought may aid us in reflecting on how hierarchies are sustained through an appeal to special capacities and the stubborn quest to maintain and elevate them. For example, during the trial, we learn that Ina know themselves to have a special capacity of reading other peoples' lies, they accept humans may have this capacity as well, but theirs is more accurate because of their especially acute senses:

> 'Of course, Ina can't sense absolute truth,' he'd explained this to me days before. 'At best, they can be fairly certain when someone fully believes what he's saying. They sense stress, changing degrees of stress. You do that yourself, don't you? You smell sweat, adrenaline, you see any hint of trembling, hear any difference in the voice or breathing or even the heartbeat. [...]. 'It's what we all do, Ina and human,' he said. 'The Ina are just a lot better at it. They do it consciously and with more acute senses. They usually have better memories, and they can pile up more years of practice than humans can.[4]

Better memories, acute senses and longer lifespan leading to more years of experience, are only some of the Ina capacities that are better than human capacities. They can sense lies, but later in the trial, they accept that this capacity is not readily available in every scenario, for instance, they are not sure they can read Shori because she is different (not entirely human or Ina):

> No one can be certain of the truth of anything you say because you are neither Ina nor human. Your scent, your reactions, your facial

> expressions, your body language – none of it is right. You say your symbiont has just died. If that were so, you would be prostrate. You would not be able to sit here telling lies and arguing. True Ina knows the pain of losing a symbiont. We are Ina. You are nothing![5]

This quote brings me to another point related to both human engineering and human exceptionalism: the anxiety of losing the purity of the species. In this case, Ina engineering and exceptionalism constantly create what an Ina is and the special capacities they harness. It creates an idea of heterogeneity, of purity that, in turn, produces a deep anxiety about losing their own Ina-ness. For conservative Ina, this anxiety involves losing something of their own way of being. Shori represents the material effect of losing one's own Ina-ness, because she is a different way of being, an impure one. Then, even if her capacity of being alert during the day, and the protection to their skin could be beneficial to Ina in general, it is not worth losing the purity of Ina species, contaminating it with human DNA.

Conservative Ina even go as far as to annihilate other fellow Ina who have been genetically intervened to protect the purity of their own species. They do not get their hands dirty though, they send humans manipulated by the exchange involved in taking blood from them. On one occasion, Shori and her friends capture one of these humans, called Victor. He discloses what other Ina told him about Shori and her family, that they were sick, making experiments on people and prostituting women and kids. He then goes on to insult her with a racist statement and she asks:

> 'They call me those things, don't they?'
> He nodded.
> 'Because I'm dark skinned?'
> 'And human,' he said. 'Ina mixed with some human or maybe human mixed with a little Ina. That is not supposed to happen. Not ever. Couldn't let you and you … your kind … your family … breed'[6]

The idea of not being a pure Ina is problematic on its own, but Victor adds to the question of how much Ina or how much human Shori might be. Is she still Ina? Is she more human? Or is she none of those things anymore. It is not only that she presents a problem with the known taxonomies and hierarchical schemes, maybe she could be something else, something different, a new species, a hybrid; but that she presents a problem of kinship.

According to sociobiological theories of kinship, the closer you might be related to any being, the more you will defend, protect or even act altruistically towards. In general, these theories, just as the Selfish Gene Theory proposed by Richard Dawkins, have been contested and to some, might have become obsolete. Different studies in queer ecology,[7] including biologist Joan Roughgarden, have shown how the competition versus altruism debate creates a false dichotomy, and that theories and experiments permeated with stereotypes, justifications and human exceptionalism (only humans can be altruistic, nature is always competitive). If anything, they show another way biological discourse may justify selfish acts of abuse and domination.

Even as throughout the story Ina insists that racism did not mean anything to them, that it was a human problem only, we see how both being dark-skinned and human, posits a problem for Ina to accept the DNA trade Shori's family has created. As the trial continues, some Ina make the point that this mix of DNA is probably beneficial to their survival as a species. The answer of one of the conservative Ina is clear as to their prejudice against black people as an extension of humans' racism: "You want to get black, human children from her. Here in the United States, even most humans will look down on them. When I came to this country, such people were kept as property, as slaves."[8]

A little bit further down the road of the trial, and linked with this past statement of the progeny that Shori and her lineage might create, the spokesperson for Ina conservative families, asks after being found guilty of planning the onslaught to Shori's family: "What will she give us all? Fur? Tails?"[9] This may show how much Ina have learned from humans about hierarchy: resorting to animal characteristics that humans and Ina don't possess (fur and tails), we are able to see how these hierarchies are placed. For Ina, mixing their DNA with humans means getting closer to animals, assuming the hierarchical human scheme where humans are superior to animals but inferior to Ina.

Surely, we might argue, this comment is blatantly racialized. As this same member of the family states that black people were slaves and property, even to humans. Racialization and animalization are tied together by history. This history of course, includes colonial history and racial justifications to legitimate slavery and genocide. Through this history, Ina have been present in different ways sometimes as slavers and in a position of whiteness that gives them concrete privileges even when they were striving to keep each other alive. Ina whiteness is particularly relevant to the narrative, as are the pureness entailed racist underlayers.

Following Sylvia Wynter's essay *1492: A New World Perspective*, the justifications of colonial enterprises throughout history changed and evolved, becoming more precise. Mostly rooted in Judeo-Christian schemes of truth where pureness is a God-given characteristic that must be procured, and at least since Illustration times, also rooted on science. These configurations varied with time but always entailed the capacity of claiming superiority and categorizing the population (as slaves and slavers) based either on a religious or allegedly scientific truth regime. Many years afterwards, biology as science, coming strongly with evolutionary theories, became a sturdy logic to justify white supremacist hierarchy, we can think of Social Darwinism[10] for example. According to Wynter, racist biology created a kind of pureness inscribed in the body, studied by physiology (phrenology) and, more recently, genetics and epigenetics.

This biological logic behind racism that Wynter identifies is also deeply intertwined with animalization processes. Following Aph Ko, this is because the biological racist discourses are tied to the human-animal dichotomy. The idea that some people (not white people) are less than human and more closely related to animals (irrational, savages, not God's people), is a founding building block to legitimizing the superiority of the white male, or Man (with capital M). This is what Aph Ko, following Claire Jim Kim, calls the zoological order of racism, or zoolo-racial order. Which follows the next logic:

> Animal as a term does not exist on it's own [...] it's relational. It only makes sense in relation to the white human. Those who deem themselves the superior humans decide who falls within the category of animal by using their own group's traits as the standard measurement. Animal is a signifier that is always convenient and changing, and any group the dominant class deems unworthy is immediately branded with this label.[11]

Which brings us back to the idea of human exceptionalism, on how a group's traits and capacities are fundamental to measure who is closer to humans than to animals in this specific hierarchical configuration, where animalization and racialization are foundational. Being in between human and animal, or in Shori's case Ina, human and animal, creates an issue of taxonomy and destabilizes the hierarchical scheme, which needs a mechanism to protect and assure the limit between the categories at play. In this sense, Ina actualize the hierarchy scheme we know, but they did not change the structure.

The narrative experiment of thinking the relationships that would be created if there were two sort of humanoid species who share a history, as well as live together in close interaction and organization, where one species (for once not the humans) have the upper hand hierarchy wise, can show in a different light how hierarchy is constructed on the idea of different capacities and how they are thought of as making one specie superior than other.

If human exceptionalism is the construction of hierarchy through capacities exclusive to humans, Ina exceptionalism is the one threatened by the DNA experiment mixing humans and Ina, that Shori was a part of.

Here, we see clearly how even a capacity that may benefit Ina, like melanin production, is not something welcomed by the conservative Ina families as the issue becomes the pureness of their own species. Ina exceptionalism creates the hierarchy and protects it by keeping a clear line between the capacities that are important to Inas' way of life and their idea of purity. As Oyèrónkẹ Oyěwùmí suggests: "Differences and hierarchy, then, are enshrined on bodies; and bodies enshrine differences and hierarchy. Hence, dualisms like nature/culture, public/private and visible/invisible are variations on the theme of male/female bodies hierarchically ordered, differentially placed in relation to power, and spatially distanced one from the other."[12]

Any profound change in understanding hierarchies' structures in society must start with taking a different approach to difference. Where difference is understood as a potential mutualistic relationship, not parasitic or based on domination and control; we may start letting go of the idea of pureness of species that creates, semiotically and materially, an impossible gap between those who have the upper hand at a certain point in time and geography and those who do not.

2. On the Breeding of Hyper-capacitated Beings, in *Wild Seed*

Wild Seed, from the *Patternist* series by Octavia E. Butler, is a story involving the technique of breeding humans with exponential capabilities. This story, aligned with the idea of uplift, shows how the transformation of life can be guided and directed by conducing the reproductive activities of individuals and populations. Unlike most uplift narratives, this story is not about creating "more human" animals, but to create more powerful humans; and the means are not necessarily scientific, but mainly a technique of trial and error that includes some fantastic elements.

The story begins with Anyanwu, a very old African woman (300–350 years old), who has shapeshifting and self-healing capacities. She is met by Doro, a man with special capacities of his own: being semi-immortal as he takes the bodies of other humans by killing them and, also able to sense other weird humans with powers of their own. Doro convinces Anyanwu of following him to the United States, where he has a community of special people like her, by promising her offspring that will outlive her. At first, Anyanwu is intrigued, but soon enough she will learn that he was not looking for companionship of someone he believed to be on his level; but for another wild seed for his large-scale artificial selection enterprise to create super-powerful human mind readers and controllers.

When Anyanwu arrives at the community, she is bred with Doro's son Isaac. She resists but soon learns submission is the only way she may remain alive, and even more, the only way to protect her actual and future offspring from Doro. Furthermore, Isaac tells her she might be the only one who may convince Doro to change, as he suggests: "Submit to him now, Anyanwu, and later, you can keep him from ever making animals of us."[13] Anyanwu tries her best to stay and take care of her offspring for years, but when she does not submit, she was punished.

Tired of Doro's constant killing, promise-breaking and cruelty, she decides to leave. She shapeshifts and finds shelter with a pod of dolphins. While living with the dolphins, she learns just how much Doro had reshaped her through the habit of submission, she promised herself she would not be his slave again. Many years pass before Doro finds her again. He believes she had been spoiled; she had known too much freedom. But when he found her, he realized Anyanwu had a colony of her own, only to her, this is not an experiment but a family. She had many people with special abilities in it too, so he decides to spare her, if he can bring his own people to mate with hers. Anyanwu knows she has no other option but to accept.

With time, Doro understands somehow that Anyanwu was the only one he could trust, he has a switch of attitude towards her and starts to treat her better. Anyanwu is not convinced by his change of attitude, she wants to be free from him and even considers taking her own life. He would not let her go, especially since she had become a gold mine for his breeding enterprise. Because, through the story, Anyanwu learns a new ability: to modify her own body, cell by cell, all the way to her reproductive gametes and genetic configuration.

This story is not a classic sf narrative in which some scientists decide to mingle with an animals' (in this case human) physiology or genes to carve a capacity from the plasticity of a body. First, because these abilities already exist in the human's Doro calls wild seeds, they can shapeshift, auto-heal, read minds and change their bodies in many possible ways. How they've come to exist is never explained, but Octavia situates Anyanwu's wild seed in Igbo's cosmology. We know this because of her use of the term *obanje*, an evil spirit, which is what she called Doro; and because of the brief history of Anyanwu's clan and the magic they possessed.

> It was said that Anyanwu's clan, the clan of her mother's husband, had members who could change their shapes, take animal forms at will, but Anyanwu had seen no such strangeness in them. It was her mother in whom she found strangeness, closeness, empathy that went beyond what could be expected between mother and daughter. She and her mother had shared a unity of spirit that actually did involve some exchange of thoughts and feelings, though they were careful not to fault this before others.[14]

In a sense, this narrative has many fantasy elements, but still some scientific strings are threaded throughout the novel. Especially because Doro's understanding of selective breeding is related to genes and gene pool, but he only knows by trial and error who to mate with whom to breed the abilities he wants in their offspring. This said, we get to understand artificial selection in two ways: as an ancient practice and a genetic understanding of heredity. For example, Doro's comment on the heredity of Anyanwu's offspring who have never had special abilities shows a basic knowledge of how dominant and recessive genes work, even when they are not explicitly named:

> 'My sons would do you no good,' she told him as they walked.
> 'They are good men, but they know very little.'
> 'They are not like you – any of them?'
> 'None.'
> 'And your daughters?'
> 'Nor them. [...]'
> 'Perhaps if your sons and daughters married each other...'
> 'Abomination!' she said with alarm. 'We're not animals here, Doro!'[15]

Wild Seed's narrative allows us to think about how selective breeding works in humans. We most often find narratives on selective breeding of animals, which is constantly present in the story as we can see on the quote above; but using selective breeding on humans, sounds like a radically problematic eugenicist practice. For Doro, this labor is seen as a construction: "I build, [...] I search the land for people who are a little different [...] I bring them together in groups, I begin to build them into a strong new people."[16]

In a profound sense, we get to understand Doro's task as a measure to create others like him, to be less alone in a world of people who do not understand him and his way of being. Even if we may feel empathy for him at some moments, it is undeniable that Doro believes himself to be superior and his trail of creating others by selective breeding is not a kind and suffering-free ordeal. Doro uses bodies to reproduce them as much as he uses bodies to continue his own life. He is not bothered by the ethical implications of using others in these ways. He excels in creating in others the habit of submission whether by discourse or by force.

He, as Anyanwu points out, is not too different from slavers of their time. He sees others as material, as producers, as objects to be used. Hortence Spiller's atomizing of the captive body comes to mind in this dynamic. The idea that slavers do not see slaves as worthy complex and complete living beings or bodies, but as fragments of something to be used: as reproductive machines, a compound of genes that may or may not have the potential to create a new powerful being. In her own words:

> This profitable 'atomizing' of the captive body provides another angle on the divided flesh: we lose any hint or suggestion of a dimension of ethics, of relatedness between human personality and its anatomical features, between one human personality and another, between human personality and cultural institutions. To that extent, the procedures adopted for the captive flesh demarcate a total objectification, as the entire captive community becomes a living laboratory.[17]

For Doro, his community is a living laboratory, to Anyanwu, her community is her family. There is a great distinction between the two. "She was not Doro, breeding people as though they were cattle, though perhaps her gathering of all these special ones, these slightly strange ones would accomplish the same purpose of his breeding. She was herself, gathering family. No doubt some of these people were of her family, her descendants."[18]

To Anyanwu her community is composed of liberated subjects, free and complete in themselves. To Doro his community is only a stock of genes he can manipulate and direct, they are, in Spiller's terms, flesh: "seared, divided, ripped-apartness."[19] This atomization of the captive body, the transfiguration into flesh, as an effect "throws in crisis all aspect of the blood relations, as captor apparently felt no obligation to acknowledge them,"[20] which is why Doro has no problem with incestuous relationships, his focus was the uplift of his offspring to create specific capacities.

The fact that in *Wild Seed*'s narrative animals are only mentioned in statements like Isaac's "stop him from making animals out of us" or Anyanwu's "breeding people like they were cattle," serves as a reminder throughout the whole novel that animals are submitted to this treatment constantly, so much so it has become naturalized. Animals caught in the industrial complex have been bred for generations to create specific capacities that are beneficial for the industries and not necessarily for the individuals. Which apparently does not create any ethical implications, because animals are already naturalized as flesh.

This process has been so profound that, as Bauer proposes, they have become postanimals (uplifted, enhanced, genetically manipulated, robotized, talking and organizing animals):

> Arguably, living pigs, cows, or chickens are always already "postanimals" in the sense of the critical posthumanist understanding of prefix I apply here: one that multiplies perspectives on human-animal binaries. Given the severe physiological changes to which centuries of 'livestock'

> breeding techniques have subjected their physiognomy, farmed animals can be understood as techno-naturecultural compounds even before cloning or other molecular breeding techniques modify them further.[21]

Critical of these interventions, Sunaura Taylor sheds light to the fact that this process of selective breeding is meant to create hyper-abled animals, which is entirely useful and productive for industries. For example, selective breeding and other technologies have created breeds of chicken, some for laying more eggs like the Leghorn breed and others to build muscle like broiler chickens. This hyper-capacitation process also comes with a disabling one.

> The issue of breeding itself raises all sorts of complex questions about normalcy, naturalness, and the boundaries between disability and enhancement. These animals are simultaneously disabled and hyperabled – made disabled by the very enhancements that make them especially profitable to industries and desirable to consumers. Disabling animals is not incidental to animal industries. It is essential for the work they do and the profit they create.[22]

These breeding issues of the animals in industrial complexes, are also important for *Wild Seed*'s narrative, in the sense that the powers Doro's offspring may get, can be also disabling to them. Some of them are unable to live in communities, cities and other large gatherings of humans, because of the constant background noise from all their mind-reading capacities. This capacity that Doro believes to be important makes their lives miserable in many ways. Even as wild seeds may resist Doro and animals resist the industry practices, they will both find new ways of keeping them from having too much freedom, even by making their resistance difficult because of the disabilities created by the enhancement process they have been subjected to.

This can be exemplified by the Sloanes family, an important part of Doro's community, their ability to feel others pain and emotions made them suffer, yet Doro would still breed them:

> When one received a burst of pain, anger, fear, any intense emotion, it was immediately transmitted to the other, and both suffered. None of this was deliberate or controlled. It simply happened. Helplessly the Sloanes did a great deal of fighting and drinking and crying and praying for it to stop happening, but it would not. […] They were amazingly good breeding stock to be wild seed. […] And as soon as they had produced a few more children, Doro intended to take them both. It would be almost a kindness.[23]

To summarize, from the selective breeding practices of *Wild Seed*, we can comprehend the way hierarchies and capacities work together to create the possibility of using other bodies to one's own benefit. As suggested earlier, this process may be tightly linked with the idea of atomization of the captive body, where the individuals are seen as flesh that can be manipulated and built upon. We can also put into question the breeding process that continues today by the means of the animal industrial complex which, in crafting postanimals, creates hyper-abled bodies for the industry and a disabled body for individuals themselves.

In this sense, we can understand how any capacity has a contextual, physiological, advantage and disadvantage; but, more importantly that directed transformation of bodies for productivity by making them hyper-able, in sf sense uplifting, has a downside and is functional only in some contexts and for some individuals but not to others. Even though selective breeding in humans is not allowed (at least without consent), this process has continued in animals, producing not only plenty of revenue, but also an enormous amount of zootechnical knowledge about genetics and reproductive practices, which could in theory be applied to humans.

Selective breeding as a directed transformation of life is present in *Wild Seed*'s narrative as being guided by a cruel and egocentric individual called Doro in his task of creating super powerful beings. The direction of these transformations is guided by the ideals and desires of the one who deems themselves in power, and the desires, necessities of those who are being transformed are not considered. In the animal industry case, direction of transformation is decided by many individuals and collectives, industry CEO's, scientists, veterinarians and government policies. Following Bauer:

> The animal industry limits the imaginability of sustainable human-animal relationships; its real-world products – that is, laying hens and broiler chickens, porkers and veal calves, as well as battery cages and disassembly lines in mass slaughter plants – are anchored as semiotic forms within our minds. A deindustrialization of the imagination entails breaking them up and, ideally, transforming them in a way that changes the perception of the living animals and the systems holding them captive.[24]

As Anyanwu faced with the immense power Doro has over her, and her impossibility of getting rid of him and his mission, animals in industries are held by the immense power of human industrial societies, that are not likely to disappear. There is only hope, as Anyanwu tells Doro that: "Someone will find a way to rid the world of you."[25]

3. On Correcting the Faulty Human Project, in *Oryx and Crake*

Oryx and Crake, the first book of *MaddAddam*'s trilogy, is a narrative by speculative fiction author Margaret Atwood. It is the story of two friends who live in a dystopic world divided into insecure, crowded and poor neighborhoods with high crime rates called Pleeblands and rich and ultra-secure industrial complexes, called Compounds. The story begins after the human species has been mostly extinguished. Snowman, apparently the only human alive, whose real name is Jimmy, presents the future made of a humanoid species named Crakers: a human animal hybrid created by his friend, Crake. Snowman and the Crakers are the only ones in the scene, apparently having survived the catastrophe that their own creator has released upon the world.

To understand how Snowman got to this point though, he reminisces about the past, letting us know that Jimmy and Crake were infancy friends, they both lived in the industrial compounds with their families, each having an accommodated, but not perfect life. They studied together at elementary school, where Crake excelled in genetic engineering practices. Jimmy, on the other hand, being the son of employees of one of the most important industrial complexes, Organic Inc. Farms, is not so thrilled about genetic engineering. As a kid, he kept being reminded by his father to not give names to the pigoons, animals from a project whose objective was to "grow an assortment of foolproof human-tissue organs in transgenic knockout pig host."[26] Pigoons were not to be companions or to be eaten but used in several other biotechnological industries such as NooSkins to develop skin-related biotechnology and even in neuro-regeneration projects, growing genuine human neocortex tissue.

Even though his mother and father wished he would be, Jimmy was not that interested in science. Crake, on the other hand, was deeply fascinated with science and became a bright student from his first experiences in elementary school on genetic engineering. Crake studies Transgenics in college, in a particularly important school called Aspergers University. He, as an autistic prodigy in science, excels in every project he is faced with. He also became obsessed with a new project of his own, an elaborate experiment to create a new species of humanoids, one that is not faulty like the *Homo sapiens*, and that has the possibility to survive and not continue devastating the world they live in.

Oryx and *Crake* was one of the narratives Bauers' used to propose the term postanimal, where she tries to understand how the modified animals in this story (pigoons, rakunks, glowing rabbits, wolvogs, bobkittens, chicken nobs and kanga-lambs) reconstitute and remake the very notion of animal and even human. In this chapter, I will focus on the idea of the faulty project that humans are, according to Crake. There are several instances in the narrative where Crake, in line with his objective scientific point of view, believes humans are deeply faulty organisms who are destroying the very planet they live and depend on.

He concocts a twofolded plan: first, the massive destruction of actual humans (The Pill project); second, the creation of a humanoid species (Paradice) that will eventually take their place and create another world from the scraps of the old devastated one.

One might read Crake's understanding of the human project faultiness as misanthropy, which could be supported by the fact that he indeed created a plan of destruction of the whole human species. Jimmy assumes that Crake, as an autistic person, is being obsessive about analyzing human faultiness. I will argue that this idea of human faultiness is more so based on his scientific understanding of life, by following Haraway's term of God-trick in science.

Crake is interested in transforming life through his understanding of biology, physiology, genetics and ecology. This God-trick is common in sf narratives involving the figure of the mad scientist, like Dr. Moreau, Dr. Frankenstein and Crake. Generally, an isolated, socially awkward and science-driven individual who finds power in the transformation of life. This is true of Crake, even though he does have Jimmy as a friend, and later in the story, many other acquaintances who help with the project.

For Crake, the God-trick comes into play in the way he believes to be capable of creating an entire new species to correct the devastation humans have created. There are several problems Crake points out about the human project, most of which he understands in a purely biological sense, in which case, genes might do the trick to correct them. For example, some are social issues tied to biological processes like reproduction. Reproductive issues are profoundly important to Crake because humans have grown in population so much they are no longer able to sustain themselves with the resources at hand. He believes this can be corrected by creating a being with signaling reproduction tactics: "Blue color abdomen and buttocks for reproduction signaling, no more unrequired love these days, no more thwarted lust; no more shadow between the desire and the act. [...] no more prostitution, no sexual abuse of children, no haggling over the price, no pimps, no sex slaves. No more rape."[27]

Others are resource-oriented issues like diet. According to Crake, this is an issue tied to the reproduction one, because on a limited resources planet there is no way to sustain a growing human population. He tells Jimmy: "*Homo sapiens* doesn't seem able to cut himself off the supply end. He's one of the few species that doesn't limit reproduction in the face of dwindling resources."[28]

More so if they are eating as much as they do with a faulty digestive system that is not using all the nutrients:

> [...] what it boiled down to was eating your own shit. [...] For animals with a diet consisting largely on unrefined plant materials, he'd pointed out, such a mechanism was necessary to break down the cellulose, and caecotrophs were enriched with Vitamin B1, and with other vitamins and minerals as well, at four or five times the level of ordinary waste material. Caecotrophs were simply a part of alimentation and digestion, a way of making maximum use of the nutrients at hand.[29]

He also adds some extra protection capacities like UV filters, insect repellant and microbe immunity capacities. As well as self-healing, which I find fun to think of as curing by purring: "Crake worked for years on the purring. Once he'd discovered that the cat family purred at the same frequency as the ultrasound used on bone fractures and skin lesions and were thus equipped with their own self-healing mechanism."[30]

Another significant one is understood as the issue of having a monkey brain: "Monkey brains, had been Crake's opinion. Monkey paws, monkey curiosity, the desire to take apart, turn inside out, smell, fondle, measure, improve, trash, discard – all hooked up to monkey brains, an advanced model of monkey brains but monkey brains all the same."[31] He also later adds: "we are hormone robots anyway, only we're faulty ones."[32] Both point to epistemic bias and hegemonic construction of discourses in science, the Cartesian view of the body-machine and the evolutionary scale of complexity that position humans at the top of all animals, including monkeys; at the same time of acknowledging the idea of being an ape, and a faulty one at that.

In this mindset, with these science constructs and in a world where genetic manipulation is common and extremely specialized, it is no surprise Crake is able to create a whole new species based on the idea of correcting the faults of human beings through the genes and potential capacities of other animals. In this God-trick situation, for Crake, genes are only parts of a mosaic structure of a body that is malleable and interchangeable. Animals and their genes are no more than a stock of potential characteristics and capacities to be mixed to create a new species to take on the world humans destroyed.

> What had been altered was nothing less than the ancient primate brain. Gone were its destructive features, the features responsible for the world's current illness. For instance, racism – or, as they referred to in

> Paradice, pseudospeciation – had been eliminated in the model, merely by switching the bonding mechanism: the Paradice people simply did not register skin color. Hierarchy could not exist among them, because they lacked the neural complex that have created it.[33]

We may believe this is a conflicting view of the world, one in which racism is only an issue of registering skin color or neuropathways to hierarchical thought, as though this was engraved in the ancient primate brain and not an effect of social histories. The genocentric way Crake applies to this project is, nowadays, problematic in its heavy biological determinism. The same issue that arises with sociobiology discourses while they were in their apex: the biological discourse infiltrates the cultural-social logic. Which can be thought of as the bio-logic in the sense of Oyèrónkẹ Oyěwùmí: "The cultural logic of Western social categories is based on an ideology of biological determinism: the conception that biology provides the rationale for the organization of the social world. Thus, this cultural logic is actually a 'bio-logic.'"[34]

Interestingly though, Crake is faced with a paradox, he believes (with scientific perspectives to prove it) that humans are a faulty project, even as he still believes they are advanced monkeys. In this sense, he is not tied to a direction of enhancing humans but destroying them; at the same time, humans are his template for Crakers. The human genome is the prevalent one, with other animals' genes mixed in to create some capacities he believes to be important in the context of the world they will inhabit. Apparently, amid this contradiction, humanness is not the goal or directive principle guiding his actions, even as it is the basis of the genetic project.

For Crake, there is no middle ground, humans must be destroyed in order for the planet to survive, but they must be replaced by a humanoid form that in some senses continues the legacy of humanness in a different way. Now, the annihilation of humans in *Oryx and Crake* is material, and the replacement is something other than human; but could we extrapolate this idea to a different, less contradicting and violent way? Could we annihilate the pure humanness discourse to give way to another form of being, one where hierarchy and superiority are not the guiding principles of transformation of life? Could we find other guiding principles like symbiogenesis and mutualistic symbiosis? Could we become another humanoid form, not human, not Crakers, but our own idea of a better species to thrive in and with the planet we share and create with other species?

In this chapter, we have revised the transformation of life while guided by hierarchy and pureness, directed to produce benefits for others and to adapt to a devastated world. BIO in this chapter has been focused on the *towards* understanding of the *into* preposition. When transformation of life is guided towards a final objective, benefit or necessity of someone; for the powerful individual but not the collective others, many issues arise. The transformation of life that becoming-into-other proposes means changing the way we live and act on this planet to create a better world for all beings today and in the future.

I agree with Bauer when she states that "literary imaginings neither aspire to yield results that are absolutely true nor make any normative claims."[35] This is important to keep in mind all the way throughout the sf chapters of this book. But what literary imaginings and, particularly, sf narratives may help us do, is pave the way by rupturing the naturalized narrative of how things are and, as Ursula K. Le Guin suggested, to train our imagination in order to create a different future.

Notes

1 Mutation is understood as the permanent hereditable change in DNA or RNA sequences created either naturally, by reproduction or because of an external factor (mutagens).
2 Speciation is the evolutionary process by which populations evolve into distinct (reproductively isolated) species.
3 Traditionally understood as the empirical investigation process which involves observation, and proposal of questions to test hypothesis.
4 Octavia E. Butler, *Fledging* (New York: Grand Central Publishing, 2005), 245.
5 Ibid., 272.
6 Ibid., 173.
7 Queer ecology refers to an interdisciplinary field which integrates queer and ecological studies. Especially there is an interest in understanding differences in sexuality and gender, by challenging the binary classical ways of explaining nature.
8 Ibid., 272.
9 Ibid., 300.
10 Social Darwinism is the use of the Natural Selection and Survival of the fittest theory proposed by Charles Darwin, to explain human societies. It has been mostly rejected, and many scholars have shown that it was used for eugenic and supremacist purposes.
11 Aph Ko, *Racism as Zoological Witchcraft: A guide to getting out* (New York: Lantern Publishing, 2021), 37.
12 Oyèrónkẹ Oyěwùmí, *The Invention of Women: Making an African sense of Western Gender Discourses.* (Minneapolis: University of Minnesota Press, 1997), 7.
13 Octavia E. Butler, *Wild Seed* (New York: Grand Central Publishing, 1980), 138.
14 Ibid., 11.
15 Ibid., 14–15.
16 Ibid., 19.
17 Hortence Spillers, "Mama's Baby, Papa's Maybe: An American Grammar Book", *Diacritics* 17, no. 2 (1987): 68.
18 Butler, *Wild Seed*, 236.
19 Spillers, "Mama's Baby…", 67.
20 Ibid., 73.

21 Liza Bauer, *Livestock and Literature: Reimagining Postanimal Companion Species* (Sheffield: Palgrave Macmillan, 2024), 18.
22 Sunaura Taylor, *Beasts of Burden: Animal and Disability Liberation* (New York: The New Press, 2017), 28.
23 Butler, *Wild Seed*, 146.
24 Bauer, *Livestock and Literature*, 16.
25 Butler, *Wild Seed*, 295.
26 Margaret Atwood, *Oryx and Crake* (New York: Anchor Books, 2004), 22.
27 Ibid., 165.
28 Ibid., 120.
29 Ibid., 159.
30 Ibid., 156.
31 Ibid., 99.
32 Ibid., 166.
33 Ibid., 305.
34 Oyèrónkẹ Oyěwùmí, *The Invention of Women: Making an African sense of Western Gender Discourses* (Minneapolis: University of Minnesota Press, 1997), IX.
35 Bauer, *Livestock and Literature*, 14.

Chapter 2

TRANSFORMATION OF LIFE IN THE AGE OF BIOPOWER

In the *Posthuman* series, hosted and produced by Emily Chang, a chapter is dedicated to the topic related to the relationship between technology and biology. *How Tech is Breaking the Rules of Biology* tells a story about science and technology intertwined and mixed in order to produce a new form of biological life. Due to scientific advances now is possible to transform any cell in the body into a gamete. But first a specialized cell (e.g. a skin cell) must become a stem cell, which is a pluripotent cell, capable of specializing in other functions, in this case in a reproductive cell. For Insoo Hyan, Director of Center for Life Sciences of Museum of Sciences, this achievement was expected in the near future, while for Amander Clark, Director of UCLA Center for Reproductive Sciences, this kind of practice had been thought only in sf stories. But these comments could be understood as rhetorical ones causing surprise to the viewers, as if the achievement had been spontaneous, random or some kind of miracle. On the contrary, it is a result of a well-designed experiment protocol, realized in a well-controlled environment known as laboratory, in which the recollection of data and the accumulation of results offer the possibility of greater success in controlling biological processes with a large number of machines, computers, instruments, tools and biotechnology. Far from being only a biotechnological and scientific achievement, this kind of practice can be understood as a biopower effect that shows that life is not only manipulated but transformed in so many forms. In this case, both sciences of life and biotechnologies exert power over life. Far from being only a future dream or simply a sf story, biopower is as present as life itself, taking it as its object. Based on this preliminary condition, that life is and exists, it is possible to instrumentalize and modify biological processes in order to produce bodies (maybe the best bodies, from an egg to human body) and reproduce life (through non-sexual contact). In Vitro Gametogenesis (IVG) is just an example of this contemporary exercise of biopower in which life is at stake. Furthermore, it is an example that shows the way in which life becomes into others by the exercise of biopower related to new scientific advances and recent biotechnologies.

However, this exercise of biopower functions and operates with some assumption of what life is. But what is life? Can life be defined? The IVG example, taken from the specific area of life sciences, only explains a biological process manipulated by biotechnologies, but does not say anything about life. Despite the lack of definition, it is stated that life is changed and transformed (from natural to unnatural) thanks to sciences and biotechnologies. But this is another assumption, that life changes and can be changed and transformed. As such, this assumption is another previous or preliminary condition that allows any subject not only to think about the change and transformation of life, but to make it real in any well-controlled environment. Finally, the life processes can be changed and transformed due to the historical birth of biopower in the nineteenth century. Within this theoretical reflection, this chapter focuses, in first place, on setting a question concerning life related with the improbable definition of it; in second place, on explaining metaphysical, materialistic and biological views about the change and transformation of life; and in third place, on understanding the contemporary exercise of biopower responsible for not only transforming life but also producing a form of life capable of BIO.

1. The Question Concerning Life

Living in biological times, where contemporary human beings consider their existence as living entities or as organic individualities pertaining to one species among others, invites us primarily to understand life in the realm of biological sciences or sciences of life. In a metaphysical view, human beings as animals are the living entities that put into question their biological existence, as well as life and its meaning. Certainly, according to evolutionary facts, there is more consensus about biological existence than about the meaning of life. First, because the concept of life is difficult to define, especially when it is used in religious, political and ethical dimensions. Second, because the meaning of life changes across history. Third, because the meaning of life has been and could be thought of in human terms. To summarize these three points, humans propose different meanings of life that could be used in religious, political and ethical dimensions in certain historical periods. With this assumption the monolithic and ahistorical perspective of the meaning of life cannot be sustained anymore.

In this impossibility to find or define only one meaning of life that runs along the whole history of planet Earth at least by human means, we could imagine some kind of possibility in which a super intelligence could answer the abysmal question: what is the meaning of life? Just like Douglas Adams has done in *The Hitchhiker's Guide to the Galaxy*, in which "hyperintelligent pandimensional beings"[1] appear interested in answering the question after a long time of vain discussions. Despite their hyperintelligence, those beings do not even try to answer the question. Perhaps their intelligence is incapable of doing so or the collective answers given are not enough. So, after the failure of thought, those beings build a super intelligent computer, the second one more intelligent in the whole universe, called Deep Thought. This computer can answer the question about the meaning of life, universe and everything, but it will take seven and a half million years to reach the answer. In this case, thanks to Deep Thought the secret or the enigma of life could be brought to light, although the answer will be incomprehensible for beings without super- or hyperintelligence (Deep Thought's answer: 42[2]). This sf story shows at least two problems in human perspective: first, the struggle for discovering and defining the meaning of life does not depend on an intelligent entity (human or even non-human but hyperintelligent) capable of doing it, but on a superintelligent entity, in an age without oracles and gods, identified with a computer and its artificial intelligence; second, human beings, incapables of answering the question concerning life, turn to machines to define it which allow to describe an image in which something that is a non-living entity could resolve the riddle of life. In other words, paradoxically, the meaning of life is only accessible to a superintelligent computer created by an inferior intelligence and, in the other case, a non-living thing knows more about life than a living being.

To simplify the object of study, the analysis should avoid these points, one that claims for a destruction of a monolithic and ahistorical perspective of the meaning of life and another one that problematizes the very access to the meaning itself, leaving aside a hidden, obscure meaning. Instead of doing that, the new object could be life itself. Then, the question concerning life should ask about life itself: what is life? But in this case, it finds another problem, because life itself does not properly have a definition. It has common and scientific definitions, but none of them are ultimate and definitive or conclusive. In this perspective, one definition gains relevance and importance only by means of the consensus in different theoretical fields, just as in medicine, biology, even chemistry and physics. Knowing that raises the question: how can the meaning of life be defined when life itself does not properly have a definition? This does not mean that, in fact, life definitions do not exist. There are a lot of definitions but every one of them represents a certain problem and difficulty, because it may be very broad allowing to include apparently living entities (every self-reproduced entity, such as fire) or very narrow to exclude different entities presumably living (if life is something organic, not every organic thing is alive). For example, Carol Cleland and Christopher Chyba, in addition to mentioning the long history of attempts beginning with Aristotle, list at least four definitions inscribed in scientific tradition whit their respective problems:

- Metabolic definition: something that consumes and converts energy to move, grow or reproduce, including every living being … fire and automobiles.
- Thermodynamic definition: a living system takes energy to create order locally. But in this case, this definition could include crystals.
- Biochemical definition: a living form is something formed by certain types of biomolecules, "yet one must worry that any such choice could in the future face exceptions in the form of systems that otherwise appear alive but are not made of our particular favored molecules,"[3] such as artificial or synthetic entities.
- Genetic/Darwinian definition: a living being is capable of evolution by natural selection, either in a chemical (excluding computers, artificial "life" and viruses) or processual perspectives (including the process by which a computer functions and life simulations). Cleland and Chyba add another objection that implies DNA replication, "that individual sexually reproducing organisms in our DNA–protein world do not themselves evolve, so that many living entities in our world are not, by the Darwinian definition, examples of life."[4]

With this variety of definitions, it becomes quite difficult to choose one that can envelop only life and living beings without failing to include, excluding or both at the same time entities that seem or resemble living beings. So, the main problem is not a lack of definitions,[5] but an absence of an ideal definition. According to Cleland and Chyba, the definition of life has to transcend, at least, the lexical (standard meanings of terms), stipulative (introduction of new meanings of terms), ostensive (use of prototypical examples to specify meaning) and operational (use of examples to specify procedures) definitions in order to reach the most informative definition able to "explain the meanings of terms by relating them to expressions that we already understand."[6] This type is known as the ideal definition. In the case of the four definitions given (metabolic, thermodynamic, biochemical and genetic/Darwinian), none of them could be considered an ideal definition, because they only "specify procedures that can be performed on something to determine whether or not it falls into the extension of the definiendum."[7] Leaving aside the lexical and stipulative definitions, the common definition of life recurs to either ostensive (this tree is alive) and stipulative (alive is something capable of evolving through Natural Selection) definitions: this tree is alive; alive is something capable of evolving through Natural Selection; then, this tree is capable of evolving through Natural Selection. For doing this syllogism that depends on a variety of meanings, first we have to know that Natural Selection is a process performed by Nature, assuming that Nature in general is alive (ostensive definition). But these definitions do not meet the criteria of an ideal definition, which seeks to specify "the meaning of a term by reference to a logical conjunction of properties."[8] This is important because it "determines the extension of the definiendum by specifying necessary [something that in its absence the term does not apply] and sufficient [something that is present in the term for which it cannot fail to apply] conditions for its application."[9] For Cleland and Chyba, the ideal definition of life is not clear, because the necessary and sufficient conditions are extremely vague (see the metabolic definition). Besides, a deepest problem emerges in relation with life and its ideal definition, since it works with categories related with human interests, but fails in provide a good definition of something related with a natural kind: "This issue is particularly important for our purposes since it seems likely (but not certain) that 'life' is a natural kind term-that whether something is living or nonliving represents an objective fact about the natural world."[10] So, life, being a natural kind term, probably has a lot of good definitions but none of them could reach the ideal type.[11] Just as Cleland and Chyba claim: "No purported definition of 'life' can provide a scientifically satisfying answer to the question 'What is life?' because no mere analysis using human concepts can reveal the nature of a world that lies beyond them."[12] If is not possible to reach the ideal definition, then it seems that the best option is to pay attention to the definition or the set of definitions used, whether is metabolic, thermodynamic, biochemical, genetic/Darwinian or another one, because it will establish the practical possibilities in relation with something presumably living and with something apparently non-living.

So now, we have three assumptions about life that seem to be a dead end, because human beings are incapable of defining the meaning of life, of accessing the realm of the meaning and of defining the concept of life itself. In the last case, the scientific approach has shown its own limits, restrictions and incapacities for making an ideal definition. However, the concept of life, apparently coined as a scientific or theoretical one and used mainly in sciences, is impossible to define. This theoretical movement nowadays is known as *definitional pessimism*, that not only claims that the whole series of definitions given across history is wrong or misconceived, but in extremis rejects any effort in defining the concept. Despite this problem the concept of life is used as business as usual, as if everyone understands (either common or theoretical sense), at least, a general definition transversal to all sciences. But as we have seen this is nothing more than an illusion or an appearance.

More radical to definitional pessimism is the historical problem that it opens. Just as Giorgio Agamben points out:

> Ivan Illich has observed that the conventional notion of life (not "a life", but "life" in general) is perceived as a "scientific fact", which has no relationship with the experience of the singular living person. It is something anonymous and generic, which can designate at times a spermatozoon, a person, a bee, a cell, a bear, an embryo. It is this "scientific fact", so generic that science has given up on defining it, that the Church has made the ultimate receptacle of the sacred and bioethics the key of its impotent foolishness. In any case, "life" today has more to do with survival than with vitality or form of life of the individual.[13]

So, in Agamben's words, life is something undefinable, general, anonymous and generic, applied and used in three dominant realms of contemporary society: science, religion and bioethics. If these realms use the concept of life without knowing what exactly life is (i.e. a cluster of bacteria in a petri dish, a foetus growing in the womb or a genetically modified organism used in agroindustry), then we can think that their practical movement is sustained in another assumption such as a principle, a principle of life.

Meaning, definition and principle: Maybe the last one provides another answer to the question concerning life. According to Eugene Tacker, the principle of life, without resorting to a single definition, implies a sort of understanding of the organization of living beings. Across Western history as Tacker claims, there are three main principles: soul, meat and pattern. Soul can be understood as something that gives life to plants, animals and humans; meat as brute matter and mechanism; and pattern as self-organization. Following these principles, we could say something about a living being: the living is something that moves because of its soul; behaves like a machine; and organizes with specific and recognizable information. These three principles allow us to describe life at the same time of describing the living. For Thacker, this trinity of principles is a triptych: "An image of thought that continuously switches, swamps, displaces and replaces the placeholder that defines life: from *psyche* to mechanism and animal electricity, to the 'gemmules' and 'pangens' to DNA and the 'code of life'."[14] Such an image is not a succession nor something like a periodical time (age-epoch). "Instead, as a trinity-triptych, they form a kind of a portrait, a face, a faciality, a field of blackholes and white walls, within which and upon which is often written: 'life is that whose essence can be deduce and yet whose essence escapes all deduction'."[15] So within this triptych, the description of life is assured but at the same time the essence (or an ideal definition) of life escapes. But not everything is lost, because the possibility of describing life and the living guarantees a field of investigation through which is able to establish and articulate boundaries: "living-non-living, organic-inorganic, animate-inanimate, but also animal-machine, human-animal, human-savage, species, races, populations, genomes…".[16] For example, Pythagoras established not to eat meat because the animal's soul probably were in another life your grandmother; vegetarian and vegan movements establish not to eat meat because animals are not machines but living and sensitive beings; scientists have created cultured meat from steam cells in laboratories without causing unnecessary suffer and pain to animals. In the three examples, meat is described as something living due to the transmigration of the soul process; the rejection of the comparison with a machine… a productive machine; and stem cell pattern replication. So, meat could be transcendent memory whose information resembles another lives and beings; an organic piece that after death lost sensation at the same time it lost its breath (psyche-soul); or a hetero-organized living thing that grows with no sensation at all, but functions just like a living machine.

So far, in relation to the question concerning life, we have found three different approaches that apparently help us in the important task of understanding life and the living. Meaning, definition and principle. But in doing so, we have reached a sort of a dead end, because the meaning of life is not accessible for human beings; the definition of life not reaching the ideal resigns itself to using ostensive (prototypical examples: this tree is alive) or operational (a tree is alive thanks to photosynthesis which is a chemical process that converts inorganic matter into organic matter) definitions or, in other words, life cannot be defined in the ideal form; and the principle of life only allows to describe life and the living letting escape the very essence of life. With a lack of meaning and ideal definition, it seems that for understanding life itself only remains the prototypical example, the process and the possible description of those. With/in this epistemological context transformation of life could be understood.

2. Transformation of Life

What do we understand when saying transformation of life? To answer this question, three main perspectives will drive the analysis to the point of transformation. These perspectives will turn around a metaphysical view of change, an extrinsic position and an immanent modification of life.

Metaphysical view of change. In *The View of Life: four metaphysical essays*, Georg Simmel states, on one hand, that life is restless by the fact that it goes beyond itself and, on the other hand, that the essence is that its transcendence is immanent. So, life can establish its own boundaries only to surpass them.

This metaphysical assumption is sustained in an anthropocentric perspective that can realize that everything has its boundaries and in some way is fixed to them. But in human perspective, this is a paradox, because humans know that they have boundaries and at the same time, know that they have no boundaries at all. In this sense, the boundary is the elemental condition for behavior and the possibility to exceed the boundary itself. Then the act of life is to be limited and to exceed the boundary. Considering human beings' spiritual life, Simmel explains that it implies a sort of self-transcendence, because in an ethical field humans overcome not only their own boundaries but themselves.

This reflection on boundaries allows us to elaborate a concept of life that transcends the human perspective in order to explain life itself. To do this, first life must be related to temporal dimension. In some sense, past and future are connected by a constant present in which life goes beyond the moment to the past and transcends the moment to the future. In this case, life is the past as well as the future. Within this temporal dimension, Simmel identifies that not only human life but animal life in general while reproducing and growing surpasses itself. If all life is capable of surpassing itself, then they are constantly going beyond themselves in the present, in continuity with the not-yet of the future.

For Simmel, another aspect of life, "as a continuous stream proceeding through successive generations,"[17] is that it flows in the form of individuals, just as a form with fixed contours. This probably seems like a contradiction for being a continuity without limits and something determined and fixed by boundaries, but this is just an appearance because while continuity cannot be cancelled life goes beyond the momentary form: "Life is at once flux without pause and yet something enclosed in its bearers and contents, formed about individualized midpoints and contrarily it is therefore always a bounded form that continually oversteps its bounds; that is, its essence."[18] That is how Simmel claims that the very essence of life is that transcendence is immanent for it. In this case, life is "at once fixed and variable; of finished shape and developing further; formed and ever breaking through its forms; persisting yet rushing onward; bounded and free; circling around in subjectivity, yet standing objectively over things and over itself."[19] Life is something that goes outside itself and fixes or establishes its own boundaries only to surpass them.

Knowing that life transcends itself, Simmel points out the last characteristic of life's essence: life is always more-life and more-than-life. The first is related with the basic condition of existence, because life could only exist as more-life, that is to say producing new things and continuously preserving its physiological dimension; while the second is related with the spiritual realm, in which something becomes a creator that produces something autonomous with specific content, proper meaning, logical cohesion and validity, capable of reaching a non-vital plane.

Extrinsic position. For Thacker, in his article titled *Biophilosophy for the 21st Century*, thinking about of a possible biophilosophy contrary to a philosophy of biology that always seeks to establish a principle of life (soul-meat-pattern) for using it to determine boundaries and limits between the living and non-living, is important to change the center of thought from the intrinsic properties of life to an extrinsic position enabling to consider that life is capable of going outside itself: "Instead of centering life (an essence, an organizing principle), what about considering life at the peripheries? Extrinsic life, a life always going outside itself, peripherical life."[20] In this train of thought, considering the extrinsic position, the articulation of boundaries runs in multiple ways not only on the living-organic path, but on the non-living-inorganic path, because life that goes outside itself intertwine and interchange with other entities, becoming-into-other and others.

Thacker claims: "Whereas the philosophy of biology is concerned with articulating a concept of 'life' that would describe the essence of life, biophilosophy is concerned with articulating those things that ceaselessly transform life."[21] Then, life does not relate only with itself, but with other entities which take the life outside itself or as Thacker asserts: "take the living outside itself."[22] More importantly, this approach questions the anthropomorphic conception of life (just as a sort of meaning underlying or preceding all aspects of living beings), that puts human life at the top of the living beings or as the great ideal living being. Nonetheless, for Thacker biophilosophy is a critique to the anthropomorphism, and it is "an approach to nonhuman, nonorganic life, anonymous life, indefinite life…",[23] that allows us to think other relationships that go beyond biological life. So, transformation of life, that is to say that living beings become into other, occurs when life enters into multiple relations with biological and non-biological life.

Immanent modification of life. In the Darwinian evolutionary perspective, life is associated with living beings that change themselves through Natural Selection or, in other words, struggling to survive and leaving offspring with modifications.[24] So, in this case, life is constant and permanent change or life transforms itself constantly and permanently during time. The set of modifications is the pattern to follow the transformation of the species across history. In these terms, the history of life in the Darwinian perspective is the meta story of the transformation of life via modifications. With/in this meta story every species, as an aspect of life, has lived on Earth constantly changing until its extinction. Just as Charles Darwin asserts in the introduction of *The Origin of Species*:

> In considering the Origin of Species, it is quite conceivable that a naturalist, reflecting on the mutual affinities of organic beings, on their embryological relations, their geographical distribution, geological succession, and other such facts, might come to the conclusion that each species had not been independently created, but had descended, like varieties, from another species. Nevertheless, such a conclusion, even if well founded, would be unsatisfactory, until it could be shown how the innumerable species inhabiting this world have been modified, so as to acquire that perfection of structure and coadaptation which must justly excites our admiration.[25]

Thus, life in the aspect of a species is constant variation and modification in the same way that a species changes throughout time. For Darwin, there is no single species that remains stable throughout its existence, because that species, on one hand, surely descended from a different one and, on other hand, its offspring generation after generation will probably but not certainly, produce another species. The fact that a single species is not created by an almighty divinity that supposedly fixes an eternal essence in every living being (and non-living too), allows us to think in a changing nature ruled by chance and regulated by complex internal–external interactions between organisms and the environment.[26] In consequence, life is not only about change but about chance and regulation.

According to these three perspectives, it is possible to assert that life changes and even more transforms itself, that is to say it goes beyond its form, whether through transcendental immanence (more-life and more-than-life), the multiplicity of relationships (life going outside itself), or generational modification (chance and regulation). Besides these theoretical approaches, the way life changes under the mode of Artificial Selection can also be analysed. After all, Darwin used this practice to understand Natural Selection,[27] as a natural and continuous process by which species change across time. But in the case of Artificial Selection, the modifications can be accumulated by humans so much faster than in nature. Just as Darwin claims: "We shall thus see that a large amount of hereditary modification is at least possible; and, what is equally or more important, we shall see how great is the power of man in accumulating by his Selection successive slight variations."[28] It is not enough only to say that Artificial Selection is related with the metaphysical dimension identified as more-life, because this practice does not correspond to the elemental condition by which life exists and continuously persists in existence. Rather it is related to the more-than-life dimension that takes the elemental condition to create other, maybe new, autonomous products that do not depend on their creator anymore. So, Artificial Selection is just a name to understand the process by which humans produce and reproduce not only life, but more-than-life that transcends physiological conservation to the point of producing another content with proper meaning. In this case, domestication as an Artificial Selection practice transcends the natural boundary[29] or the momentary fixed and established form of a living being. If life transcends itself via Natural Selection, then human beings transcend life and themselves by Artificial Selection. A life driven beyond and outside itself by Artificial Selection confirms the fact that all living beings are interconnected and intertwined with other entities (organic–inorganic, living–non-living) interchanging substances and materials. Artificial Selection confirms the hereditary process of Natural Selection, while confirming that life transforms itself and can be transformed for other means that transcend the biological boundaries and means. This process of transforming life into others has received the name of biopower.

3. Transformation of Life in Biopower Era

> The key is man's power of accumulative selection: nature gives successive variations; man adds them up in certain directions useful to him.
>
> —Charles Darwin, The Origin of Species

"The great power of this principle of selection is not hypothetical."[30] But its power across history has not been the same. In order to point out that greatness of this kind of power, and to account for the specificity of the practice in his present, Darwin quotes two statements given by "highly competent authorities"[31]: Youatt who affirmed that selection "not only modify the character of his flock, but to change altogether. It is the magician's wand, by means of which he may summon into life whatever form and mould he pleases"[32]; Sir John Sebright, in relation to pigeons, who "would produce any given feather in three years, but it would take him six years to obtain head and beak."[33] These statements show the exercise of power over some animals (cattle or pigeons) to change altogether the organization of the flock's life and the materiality of animal's bodies. So, through this principle of selection, humans exert a form of power that seeks to produce life and bodies by controlling biological reproductive processes. In this case, Darwin explains that the principle of selection is not a modern discovery, because the breeding of domestic animals has been practiced and registered for humans since antiquity. But that historical reference does not account for the specificity and discontinuity of the principle of selection itself. Precisely, this new form produces changes by another means: "I am convinced that the accumulative action of Selection, whether applied methodically and more quickly, or unconsciously and more slowly, but more efficiently, is by far the predominant Power."[34] But where does the predominance of this exercise of power lie?

The effectiveness of power does not lie in new breeding practices or techniques based on different methodologies, unknown in antiquity or another time in human history. Rather, and more exactly, this exercise of power over life, thanks to the "threshold of a modernity that we have not yet left behind,"[35] situates in its core a historical episteme that conditions the production of knowledge. In this case, according to Michel Foucault, the historical episteme sets a quasi-transcendental known as life that allows to produce a certain kind of knowledge of life itself. At the same time, this quasi-transcendental produces its objects, the methods, instruments, and tools for knowing the living beings and their processes by which they are and remain alive. This episteme introduces history within life at the same time as allows us to understand its history through and across temporal changes. Just as Foucault claims: "Historicity, then, has been introduced into nature – or rather into the realm of living beings, but it exists there as much more than a probable form of succession; it constitutes a sort of fundamental mode of being."[36] For Foucault, one of its consequences is that "life is expressed in the form of animality,"[37] at the same time that the animal "discovers fantastic new powers in the nineteenth century."[38] More important in his historical view is the transition from the images of plants to animals:

> But when characters and structures are arranged in vertical steps towards life – that sovereign vanishing-point, indefinitely distant but constituent – then it is the animal that becomes the privileged form, with its hidden structures, its buried organs, so many invisible functions, and that distant force, at the foundation of its being, which keeps it alive. If living beings are a classification, the plant is best able to express its limpid essence; but if they are a manifestation of life, the animal is better equipped to make its enigma perceptible.[39]

According to Foucault's analysis, the form of animality takes possession of the place that allows us to perceive the manifestation of life. In biology, the form of the plant is no longer essential to denote life, but rather that of the animal, since it is precisely the image of the hidden, the covered, the invisible and the distant force, whose power is always in relation to other elements. It is also possible to conclude that if one of the consequences for the West of life entering history was the primacy of the form of animality, then it must note another great consequence that is no longer in the order of the approaches of biology, but within political order. Here, precisely with the form of animality, it is possible to configure a biopolitics that, when thinking about life, does so opportunely from the form and values of the animal. Foucault highlights the primacy of the form of animality, which makes it suggestive to point out that the biopolitics that emerged during the eighteenth century were directed towards the regions of the hidden, the covered, the invisible and that distant force that lies in the body of the species, that is, in the population. Consequently, the population is associated with the form of animality. Thus, Foucault recognizes the nineteenth century as the threshold of modernity, when life had not only been introduced into history in the form of animality, but also "life of the species is wagered on its political strategies."[40] This new threshold precisely marks the beginning of "an era of 'biopower',"[41] which implies both technologies of power, on one hand, an anatomopolitics of the human body "centered on the body as a machine: its disciplining, the optimization of its capabilities, the extortion of its forces, the parallel of its usefulness and its docility, its integration into systems of efficient and economic controls, all this was ensured by the procedures of power that characterized the *disciplines*"[42] and, on other hand, a biopolitics of the population "focused on the species body, the body imbued with the mechanics of life and serving as the basis of the biological processes: propagation, births and mortality, the level of health, life expectancy and longevity, with all the conditions that can cause these to vary. Their supervision was carried out through an entire series of interventions and *regulatory controls*."[43]

Configured from the form of animality, individual bodies and population are subjected to the calculation and management of their biological life: functions, processes, relations with the environment, care and promotion of their conditions of existence. All this follows from the fact that humans, as living beings, present regularities in their functions or in their biological processes, which can, moreover, be framed in a global project of normalization and regularization.

Finally, through Foucault's analyses of life and history, the figure and form of animality that characterizes life from the nineteenth century onwards has been brought to light. One of the consequences of this event is that the life of individuals and populations can be perceived in the form of animality; in effect, the population can be animalized to the extent that it can be understood from its biological aspect. In this sense, both the body of the individual and that of the population belong to a species. Now, it is possible to think that the identification of life with the form of animality silently accompanies

Foucault's later studies in which he refers to biohistory and biopolitics: "If one can apply the term bio-history to the pressures through which the movements of life and the processes of history interfere with one another, one would have to speak of bio-power to designate what brought life and its mechanisms into the realm of explicit calculations and made knowledge-power an agent of transformation of human life."[44] Likewise, it can be thought that life across biohistory and in any biopolitical strategy is susceptible to being analysed from the aspect of animality. But what is at stake is the transformation of human life by biopower. This is the fundamental characteristic of the threshold of biological modernity from which, as Foucault claims, we have not left behind and which, therefore, still determines life in the West.

Through Foucault's analysis, it can be understood that Darwin's assumptions and statements on modern Selection were not based on a radical change in breeding methods. Belonging to the nineteenth century, to the threshold of modernity, Darwin, just as the methodic breeding of domestic animals, is just an effect of, on one hand, an episteme that introduces history into life and, on the other hand, an exercise of power over life. So, the Principle of Selection in the modern version is conditioned by the quasi-transcendental known as life by which a subject can obtain certain knowledge, a biological one, of living beings as objects centered in animality or in animal form. This knowledge, being classified, produces another form of selecting animals.[45]

In this case, can it be said that the modern Principle of Selection becomes a form to exert biopower? Foucault's analysis addresses only the modes in which power is exerted over human life whether on the individual body (anatomopolitics) or on the population life (biopolitics), that is to say that power, in a more productive way, produces bodies and life. But if one considers that selection aims to produce animal bodies (the best bodies) and life (reproduction through generations), then it is possible to affirm that selection is another technology of power centered in breeding domestic animals to produce the best bodies and reproduce life. But more radical, what if anatomopolitics and biopolitics, just as Principle of Selection, aim for methodical breeding of humans, controlling and regulating biological processes in order to produce bodies and reproduce life? The answer could be, in a non-anthropocentric sense, that biopower transforms human life, just as it transforms domesticated animal life too or, in an anthropocentric sense, that biopower transforms animal life just as it transforms human life. Both senses open a gap where human and animal life become indistinct for the exercise of biopower.

At first glance, this partial conclusion may seem strange and somewhat ridiculous by the fact that humans, in a humanist frame, do not select themselves as if they were animals. But in the twenty-first century, this assumption has come to another stage where a new form of Selection based, above all, in genetic biotechnologies is not only used in different animals or plants, but in human beings themselves. This Neo-Darwinism, as some are calling this scientific movement, "holds that selection only occurs at the genetic level, in the sense that variations acquired at the somatic level are not passed from one generation to the next."[46] In consequence, following the modern principle of selection, the scientific and medical practices and interventions have to be centered in the modification of genetic material or, in a more specific sense, have to select the best genetic material for producing human beings whose physical, cognitive and affective conditions become better and greater. This new type of selection will produce, as selection in animals, the best bodies and another kind of life reproduced through non-sexual biotechnological processes and procedures. This new type of selection shows that in the nineteenth-century animal selection, as a form of biopower, was intertwined with other technologies of power, anatomopolitics and biopolitics, producers of bodies and reproducers of life. At this point, it is not so strange that in recent days, thanks to Neo-Darwinism or Neo-Selectionism and so-called liberal eugenics, the new practices are related with the principle of selection but, in a new expression, as a biopolitical selection of human beings. The change of century has shown a different form of the exercise of biopower, focused on transforming life while transforming the genetics of living beings.[47] Several examples can be found, just as transgenic corn, cloning of dogs and horses, the production of stem cells, the testing of pre-implantation stage embryos or human design by modifying their genomes, but recently there was a case that linked and connected in different way the exercise of biopower with the principle of genetic selection. In 2022, Bartley Griffith performed the first genetically modified heart transplant.[48] The heart, coming from a pig produced by a genetic edition, is contemporary evidence that the exercise of biopower makes life indistinct. But in this case, it needs the inorganic and non-living material of biotechnologies to achieve it. Moreover, if life is transcendent and extrinsic *per se*, then biopower produces a form of life that transcends and goes beyond life itself. In consequence, life becomes into others when it is transformed by biopower technologies.

So far, the way in which transformation of life takes place has been explored and, in a deeper sense, the way in which life becomes into others. Leaving aside an ideal definition of life allows us to focus on the description of the processes by which life changes across time. In the metaphysical view, life is always more-life and more-than-life, because it always transcends itself whether in a biological or spiritual form. This form implies that life transcends itself, even at the edge reaching death or non-living things (e.g. a book or a wooden table). For life, transcendence is immanent, then life always becomes into others. In a more materialistic view, life goes out itself, because it is extrinsic, that is to say that relationships established with other forms of life and with other non-living and non-organic matter change living beings. So, life becomes into others when relating with other living, non-living, organic and non-organic matters. In a biological view, living beings as aspects of life change across history through successive variations and modifications produced by a transgenerational reproduction process. Then, life becomes into others when a new species or race emerges in nature. Further, the Principle of Selection, as a technology of biopower, has shown another mode in which life is transformed in the same way that bodies and life are produced through anatomopolitical and biopolitical technologies, as much as with genetic biotechnologies. In this case, biopower transforms life, controlling biological processes, while it produces a form of life that transcends and goes out and beyond itself. Life becomes into others in so many ways in the Age of Biopower.

Notes

1 Douglas Adams, *The Hitchhicker's Guide to the Galaxy* (London: Gollancz, 2002), 134.
2 Ibid., 146.
3 Carol Cleland and Christopher Chyba, "Does 'life' have a definition?" in *The Nature of Life. Classical and Contemporary Perspectives from Philosophy and Sciences*, Ed. Mark A. Bedau and Carol E. Cleland (Cambridge: Cambridge University Press, 2010), 327.
4 Ibid., 328.
5 To know other problems of the definition of life, see the next list made by Carlos Mariscal and W. Ford Doolittle: "Life is organized. So are geological formations. Life processes energy, so does fire. Life evolves, so do star system. Life is at thermodynamic disequilibrium, so is a car. Life is made of complex biochemistry, so are prions. Life exhibits functional complexity, so does a clock. Living creatures embody contained, functionally integrated programs, but so do robots. Life reproduces. Mules don't, bubbles do. Life is self-sustaining. Parasites are not, the carbon cycle is" (Carlos Mariscal & W. Ford Doolittle "Life and life only: a radical alternative to life definitionism", *Synthesis*, vol. 197, (2020), 2977-2978, https://doi.org/10.1007/s11229-018-1852-2).
6 Cleland and Chyba, "Does 'life' have a definition?", 329.
7 Ibidem.
8 Ibid., 329-330.
9 Ibidem.
10 Ibidem.
11 Edouard Machery explores this topic in the article *Why I stopped worrying about the definition of life… and why you should as well*, where he states emphatically that defining life is a misconceived project, either impossible via using folk concepts that in fact they are not properly definitions (this tree is alive) or pointless via using theoretical definitions cause they are so many that is impossible to choose one (i. e. choosing between the metabolic or biochemical definitions).
12 Ibid., 332.

13 Giorgio Agamben, *The Use of Bodies. Homo Sacer IV, 2* (Stanford: Stanford University Press, 2016), XX.

14 Eugene Thacker, "Biophilosophy for the 21st Century" in *Resisting Bio-politics. Philosophical, Political, and Performative Strategies*, edited by S. E. Wilmer and Audronė Žukauskaitė (New York: Routledge, 2016), 123.

15 Ibidem.

16 Ibid., 124.

17 Georg Simmel, *The View of Life. Four Metaphysical Essays* (Chicago: The Chicago University Press, 2010), 9.

18 Ibidem.

19 Ibid., 10.

20 Thacker, *Op. Cit.*, 125.

21 Ibid., 126.

22 Ibidem.

23 Ibid., 127.

24 At the end of the introduction of *The Origin of Species*, Darwin claims that "I am convinced than Natural Selection has been the main but not exclusive means of modification" (Charles Darwin, *The Origin of Species* (Hertfordshire: Wordsworth Editions Limited, 1998) 7).

25 Ibid., 4.

26 Tacker explains the complex internal-external interaction with the differentiation established as inward-outward turning: "The inward-turning aspect is metabolic, in that it processes, filters, and differentiates itself internally; it is the breakdown and production of biomolecules, the organization of the organs, the genesis of the species and races. The outward-turning aspect is immunologic, for it manages boundaries, exchanges, passages; it is the self-nonself distinction, the organism exchanging with its environment, sensing its milieu, the individual body living in proximity to other bodies" (Thacker *Op. Cit.*, 124).

27 Contrary to other naturalists, Darwin was interested in the studies of domesticated animals and cultivated plants, because, in order to understand modifications and coadaptation, they offer "the best chance of making out this obscure problem. Nor have I been disappointed; in this and in all other perplexing cases I have invariably found that our knowledge, imperfect though it be, of variation under domestication, afforded the best and safest clues" (Darwin, *Op. Cit.*, 5).

28 Ibidem.

29 Before Darwin, J. B. Lamarck was conscious of this effect produced by domestication practices. In *Zoological Philosophy* pointed out: "Where in nature do we find our cabbages, lettuces, etc., in the same state as in our kitchen garden? And is not the case the same with regard to many animals which have been altered or greatly modified by domestication?

How many different races of our domestic fowls and pigeons have we obtained by rearing them in various environments and different countries; birds which we should now vainly seek in nature?" (Jean-Baptiste Lamarck, *Zoological Philosophy. An Exposition with Regard to the Natural History of Animals* (Chicago: The University Chicago Press, 1984), 110).

30 Darwin, *Op. Cit.*, 26.

31 Ibidem.

32 Ibidem.

33 Ibidem.
34 Ibid., 35.
35 Michel Foucault, *The Order of Things: An Archaeology of the Human Sciences* (London: Routledge, 2005), xxvi.
36 Ibid., 300.
37 Ibid., 301.
38 Ibidem.
39 Ibid., 302.
40 Michel Foucault, *The History of Sexuality: an introduction* (New York: Pantheon Books, 1978), 143.
41 Ibid., 140.
42 Ibid., 139.
43 Idem.
44 Ibid., 143.
45 Beyond statements, Darwin includes a description of a modern breeding practice that enables to understand the effect of the historical episteme: "In Saxony the importance of the principle of selection in regard to merino sheep are placed on a table and are studied, like a picture by a connoisseur; this is done three times at intervals of months, and the sheep are each time marked and classed, so that the very best may ultimately be selected for breeding" (Darwin, *Op. Cit.*, 26).
46 Nathan Van Camp, *Redesigning Life. Eugenics, Bio-politics, and the Challenge of the Techno-Human Condition* (Bruxelles: Peter Lang, 2015), 18-19.
47 In *The Crisis of Medicine or the Crisis of Antimedicine?*, Foucault notes that in the second half of twentieth century, human beings had entered into a new phase in biohistory when they made genetic modification possible: "Nowadays, with the techniques at the disposal of medicine, the possibility for modifying the genetic cell structure not only affects the individual or his descendants but the entire human race. Every aspect of life now becomes the subject of medical intervention. We do no know yet whether man is capable of fabricating a living being which will make it possible to modify the entire history of life and the future of life (Michel Foucault, "The Crisis of Medicine or the Crisis of Antimedicine?", *Foucault Studies* vol. 1, (2004), 11, https://doi.org/10.22439/fs.v0i1.562).
48 The patient, not being a candidate for a human heart transplant, decided to undergo surgery to increase his life expectancy. This case confirms what Boris Groys understands as the great achievement of biopolitical technologies focused in "the shaping of the lifespan itself" (Boris Groys, "Art in the Age of Bio-politics: From Artwork to Art Documentation" in *Art Power* (Cambridge: The Mit Press, 2008), 54). On one hand, transplant seeks to extend human life but, on other hand, reduces the animal lifespan. That's why Groys claims that biopolitics not only produces bodies under design.

Part II

EXTINCTION OF LIFE OR MOVING TOWARDS PRACTICES OF CARE AND BIOPOETICS

Chapter 3

EXTINCTION OR TWISTING TOWARDS RENEWAL OF LIFE ON EARTH

We live on a planet that seems to be in a constant state of emergency: global warming is already having several effects on different parts of the world, species numbers continue to plummet, air pollution is on the rise, the oceans and other water bodies are being polluted, forests are being brought down by millions of acres, either by fires or animal industries settling in. In México, we are facing the catastrophe of several megaprojects and industry takeovers of land, truly modern colonization projects, with an undoubtedly several negative environmental, health, social and cultural impacts. Impacts are hard to grasp even in their short-term consequences.

While writing this chapter, in the north of Mexico, there is a project called Saguaro for a massive introduction of gas pipelines that will be constructed right on the Gulf of Cortes. A gulf that Jacques Cousteau once called the Aquarium of the world due to it being the home to many species of marine mammals, fish, corals, invertebrates, reptiles, algae and birds. There is a legitimate concern among several groups regarding the opacity of environmental mitigation of negative effects on such a rich ecosystem. And this, obviously, is only one example among several around the planet.

In the face of the global era of extinction, not only of species and populations, but of entire ecosystems, there has been many volumes of academic literature devoted to understand humans involvement in the several environmental disablings, as well as many creative or fantastic literature (sf and many others) to keep on trying to understand the complex problems behind this devastation; as well as what, if anything, we could do to remediate or live with the problem whether with grand solutions or at least in very local and situated ways.

From a feminist standpoint, ecofeminism was one of the first proposals (situated in the 1970's) focusing on showing the relationships between the systematic devaluation of nature (ecosystems and beings) with the equally systematic devaluation of women or the feminine. As a critical example, we can mention François d'Eaubonne, a student of Simone de Beauvoir. To this day, ecofeminism is a plural field, as much as feminisms in general, but always emphasizing the need to comprehend this dual devaluation of nature and women,[1] while also understanding that depending on the social setting, people suffer in different ways the effects of climate change and pollution, and that there are, as well, differentiated responsibilities and practices of care in the face of these environmental problems. In part, because the gendered constructions that exist imply heavier workloads of chores of care for women than for men, and because the environments we live in may be more or less vulnerable or affected, and that in the political/structural systems in which we live and the possibilities of survival we have access to are never the same for everyone.

This means that, even if we are not equally responsible for the tremendous global scale of environmental destruction, we all partake in the world we inherited in our own possibilities. And this spectrum of possibilities could be understood as part of the construction of a system that Haraway calls White Capitalist Patriarchy (or the scandalous thing) which "turns everything into a resource for appropriation."[2] This systematic dynamic of appropriation is interconnected with several other material and semiotic constructions of hierarchy: sexism, cis-heteronormative, classism, speciesism and racism. What are we to do when faced with the complex phenomenon of extinction enmeshed in the scandalous thing? How is the scandalous thing related to extinction? And who is at stake in the processes of extinction?

When referring to extinction, at least in the biological sciences field, the term refers to one level of organization: species. The extinction of a species though, has several impacts on the whole ecosystem, on populations and communities. Ecosystems are a complex web of interrelations, if one piece is lost or is removed, the whole ecosystem might collapse. A known and paradigmatic example is the Yellowstone Park wolf's local extinction and their reintroduction. We, as collective humans, especially those surrounded by techno complexes we call cities, feel exempt from this interdependence, but this is just a fictional tale we tell ourselves. Even surrounded by our commodities, we are part of this world of interrelations.

It is possible to find some extreme ideas regarding the future of the planet and life on it. For example, the argument that we as humans are not going to annihilate all life on Earth; some bacteria and fungi will survive even in the most polluted, heated and unstable environment. Even if this is probably true, this argument seems to forget the fact that many species and many lives will be lost and are being lost, as much as it covers up our personal and collective accountabilities for these deathly events. It calls for passiveness, for a naturalized perspective about extinction and a pessimistic attitude that prevents us from seeing what we do in this world matters. What then, are we to do with the complexity of extinction processes? How do we hold each other accountable and how do we become less deathly? Should we focus on rewilding[3] and rehabilitating ecosystems? Or is it that the solution is to keep a genetic blueprint of species to conserve biodiversity?

Rossi Braidotti calls for finding the faces and cartographies of extinction. As I understand it, this means not only to recognize and value the lives that are lost during extinction events as much as being able to recognize the structures of power that created the extinction chaos in the first place. In her words: "Cartographies aim at epistemic and ethical accountability by unveiling the power locations which structure our subject-position. As such, they account for one's locations in terms of both space (geo-political or ecological dimension) and time (historical and genealogical dimension)."[4]

The power locations are connected to the scandalous thing, which for Haraway are fundamental to understanding the ways we can reconfigure our relationships with others in the Anthropocene and Capitalocene. Haraway believes that extinction is a slow process of transformation of relationships in which the imaginaries of the term species play a fundamental role. These imaginaries "structures conservation and environmental discourses, with their 'endangered species' that function simultaneously to locate value and to evoke death and extinction in ways familiar in colonial representations of the always vanishing indigene. The discursive tie between the colonized, the enslaved, the noncitizen and the animal – all reduced to type, all Others to rational man, and all essential to his bright constitution – is at the heart of racism and flourishes, lethally, in the entrails of humanism."[5]

De-extinction processes have been, as of lately and at least discursively, at the centre stage as a new conservation practice. The idea of bringing back species that were once extinct has become a reality instead of an sf story in the flesh and bones of the Dire wolf puppies. But if we are to think about the cartographies of de-extinction, of the geopolitical and ecological dimensions of this process and its ties with the species discourse, many questions arise. Is then de-extinction a process of revivification to maintain, conserve the apparent purity of a past to avoid the slow process of transformation of relationships that extinction means? What does the de-extinct puppies who were brought to life mean to the constitution of the bright constitution of the rational man? Can de-extinction occur without a land full of diverse others, in constant relationship with one another?

Throughout this chapter, I will be referencing Sunaura Taylor's book *Disabled Ecologies: Lessons from a Wounded Desert*, in which she suggests we live on The Age of Disability,[6] where ecosystems as well as people become disabled by the intervention of industries in the world. As ecosystems are intervened and made disabled in their own ways, disrupted cycles and mechanisms, like water systems being polluted by armament industries, people and beings living in the land are affected corporally by this pollution and may become disabled themselves.

This Age of Disability is not an addition to the Anthropocene era, or the Sixth extinction era,[7] but a way of naming a profound interconnection of the lived and materialized effects of the actual pollution of the world. In this sense, her proposal of a disabled ecologies framework is an invitation to think about the relationships between environment and beings' (including humans) health and wellbeing. Taylor requests us to look deeply into the profound connections we have with the world and, from there, to try and find ways to heal together.

Of course, there are many initiatives of rewilding and conservation making supreme efforts to protect, not only endangered animals for the sake of conservation, but also to enable a territory for them to thrive, as well as practices to aid them in the adaptations they need to live on the new or renewed land. Nevertheless, it seems necessary to take one step further. In effect, there's been also growing efforts to reimagine and recreate land division and ownership to include animals, plants and whole ecosystems' right to land and territory. Two proposals I would like to highlight are *Zoopolis* by Sue Donaldson and Will Kymlicka, as well as *Wildlife as Property Owners* by Karen Bradshaw. These projects are, in my perspective, necessary and a great step into remaking the world we live in to increase our and other's possibilities of survival. As well as, I believe, they could be fortified by discussions brought to light by the disabled ecology framework.

A disabled ecology framework focuses on the story and process of disablement (injury, pollution, intoxication, devastation) for two reasons: first to shine light on who is accountable, under what systems of power or what cartographies of oppression are at play and, second, to propose mitigating, supporting and healing strategies for all who have been affected. As such, in this chapter, I will focus on thinking with the framework of disabled ecologies and stories of extinction written by two different authors of sf narratives.

In this chapter, I try to comprehend better how extinction can be understood as a process of relational transformation of the world, environment, territory, time and beings. As a slow or rapid reconfiguration of relationships, loss and creation of ways of being tied to a place or world, to a past and a future and to others. It is a trajectory, a voyage, a cartography, a web of relationships and individual lives being reassembled. At the same time, extinction processes guided by humans through the disabling of the environment, populations and communities, in which it is necessary to understand who is accountable, as well as to create mitigating, supporting and healing strategies.

The chapter is divided into two sections. During the first one, I will focus on a parallel play of the story of the alien species who saved humans from the nuclear war they perpetuated, with a price, through the story of *Dawn* by Octavia E. Butler; and the conservation practices based on genetic conservation and artificial reproductive technologies in the paradigmatic case of Northern White Rhinos. On the second section, the accidental survival of chickens to a polluted ecosystem thanks to their adaptations to a rough (to say the least) environment of the Barn Universe, in *Barn 8* by Deb Olin Unferth, parallel to an also paradigmatic case of wildlife survival in radioactive grounds, the dogs of Chernobyl.

1. On Conservation Efforts, the Right to Be Rescued and How to Be Treated, in *Dawn*

Octavia E. Butler's novel, *Dawn*, is the first book of a saga called *Xenogenesis* in which, after a huge nuclear war on Earth, an alien species called Oankali take some remaining humans into their ship while they restore the devastated planet. *Dawn* is the story of Lilith, the human who has been selected to be the guide of humans back to their home planet. The story takes place onboard the Oankali ship, as a space-time limbo in between past Earth and future renewed Earth, amid a transformation of the planet as well as the continuing humans.

Lilith is the first human to survive being awakened from the induced sleep she was put on by the Oankali. She soon asks them if there is a price to pay for their aid, the Oankali tell her they understand it as a trade. They are a trading species who roam the universe and trade with other species by learning from their bodies, being experts in DNA recombination. The ship they are in is also evidence of how much they have advanced this knowledge, the ship is alive, it is a creature, they feed it, control it and relate to it in different ways. This process of trading is not a unidirectional relationship: "We trade the essence of ourselves. Our genetic material for yours."[8] As travelers, they are a species not confined to their home world; they believe the original world from where they were possible: "[…] was a womb. The time had come for us to be born."[9]

Lilith learns that Earth has been importantly modified, it is no longer the world she knew growing up. The Oankali inform her that "between the efforts of your people to destroy it and ours to restore it, it has changed."[10] It was now: "Wild. Forests, mountains, deserts, plains, great oceans. It's a rich world, clean of dangerous radiation in most places. The great diversity of animal life is in the seas, but there are a number of small animals thriving on land: insects, worms, amphibians, reptiles, small mammals."[11] If humans are to survive, they must learn to live with this new wilderness.

Lilith is constantly suspicious of the Oankali treatment and motives, she understands that she's been awakened to help others go back to Earth, but she feels, instead of grateful for her saviors, like a guinea pig in an elaborate experiment. This idea is an important one, as she references it in many points of the story. She believes she and the other humans are being treated like animals, or better yet, how humans treated animals in past Earth and, as such, she feels their humanity is being taken away from them. What humanity means varies amongst the characters, but to Lilith, one important characteristic of humanity is her free will, her autonomy or in a sense, the fact that she does not feel like she can decide how she wants to live or decide her own faith.

In this sense, this story may align well with the idea of stories of disabled ecologies, in two ways. First, it exposes how systems of power are configured in the old human Earth society, and what happens when confronted with a new relationship of a society of different understandings of power. Specifically, the hierarchy between humans and animals. Second, by exposing these systems and the ethical attunements that they produce and reproduce, it becomes "useful for thinking through what kinds of assistance, care, and assess harmed people and environments need in the long aftermath of injury."[12]

For the Oankali, it is clear that human characteristics that made them almost extinct as a species were their intelligence (which they value and compliment) and their hierarchical configurations. Regarding hierarchy they believe: "That's the older and more entrenched characteristic. We saw it in your closest animal relatives and in your most distant ones. It's a terrestrial characteristic. When human intelligence served it instead of guiding it, when human intelligence didn't even acknowledge it as a problem, but took pride in it or did not notice it at all…"[13]

Let us try to read this without applying the dichotomy of nature–nurture or biological–cultural we are used to. The Oankali are expert traders, they tend to understand life as relationships, they value change, transformation, connection and combination. They understand hierarchy in humans as a mechanism of defense confronted with others, "Different is threatening for most species, […]. Different is dangerous. It might kill you."[14] But to Oankali different is necessary, valuable, important, it is the whole point of their travels and their species. They are not thinking about what the other species might want, or how much their dependence on them will affect their judgement; they believe what they can give (themselves) is enough, is valuable, and humans will come to terms with the trade.

But, to Lilith this is experienced as a radically unequal relationship. She resists: "No! […] A rebirth for us can only happen if you let us alone! Let us begin again on our own."[15] She felt as: "She was intended to live and reproduce, not to die. Experimental animal, parent to domestic animals?

Or… nearly extinct animal, part of a captive breeding program?"[16] She eventually gets to know, she wasn't entirely wrong, Oankali have messed with her reproductive capacities, they gave all humans what "[y]our people call it birth control."[17]

Even though they explicitly explain that her reproductive abilities can be restored, and they will when she is ready, she still believes this to be an intrusive and non-consensual procedure. It could be argued that in a nonhierarchical society, the idea of experimental as seen in a relationship of power over, subject–object, captor–captive, experimenter–experimented on, could be dramatically different, or even not perceived as a superiority and control. But still the Oankali are messing with bodies of other species in a way that is intrusive and non-consensual. They are the ones in charge of human's future; they are the ones with information and the ones who can transform and control.

Lilith starts to come to terms with the Oankali as she is paired with Nikanj, an infant Ooloi, a type of Oankali neither male nor female, with exponential capacities of modifying other beings' DNA. Even as she still feels suspicious of the Oankali, she becomes trusting on Nikanj because of their honesty, vulnerability and juvenile naivety. This is why, before she embarks on the mission of awakening humans, she agrees to be transformed by Nikanj, making her stronger, more agile and giving her self-healing properties, to better deal with the task ahead.

As Lilith starts awakening humans, problems arise. Many do not believe in her and are suspicious of everything she says, others become enraged and believe they can escape, and others believe she is, herself, part of the aliens. Alliances and enemies are created, and Lilith struggles to bring them together. She, having spent more time with the Oankali, has accepted the truth of their reality and the impossibility of escape without help of the alien species who 'saved' them: "We're endangered species – almost extinct. If we are going to survive, we need protection."[18]

The different events that occur after the awakening of humans show how much the narrative of feeling like humanity is being taken away from them impacts them. In Lilith's view, while saving a woman from being raped by a man, they are to remain human: "We stay human. We treat each other like people, and we get through this like people. Anyone who wants to be something less will have his chance in the forest. There'll be plenty of room for him to run away and play at being ape."[19]

When the Oankali present themselves to the group for the first time, they drug the humans to make them calm during the encounter, especially because they are to create groups of Oankali and humans to bond. One of the humans resists, he's afraid that "[h]is humanity was profaned. His manhood was taken away."[20] He tries to kill one of his bonding companions and is instantly killed by one Oankali's tentacle stinger. This occurrence did not make the other humans afraid, it actually made them feel the man was right in rebelling: "He died human! And he almost managed to take one of them with him!"[21]

After this incident, humans are taken to the training grounds, a huge room made to look, feel and smell like a jungle. But humans did not believe they were still on the ship; they started fleeing and wandering into the wilderness. Lilith is pressured to accompany them, even though she knows there is nowhere to escape to, and that their time would be better invested in learning to survive than trying to rebel. When away from the Oankali, Lilith's group gets attacked and her partner Joseph is killed. Mayhem starts as the Oankali intervene and fight back. Nikanj is injured and Lilith lies with them to cure them in the middle of the battlefield.

A while after, when humans are about to be sent back to Earth, Lilith learns they are not going alone, as one Oankali tells her: "They need us now. They won't have children without us. Human sperm and egg will not unite without us."[22] She also learns that she has been made pregnant with Joseph's child, by Nikanj's intervention, she is shocked and reproves this intervention on her body. Their final conversation shows how much the trade will involve:

> 'Our children will be better than either of us,' it continued. 'We will moderate your hierarchical problems and you will lessen our physical limitations. Our children won't destroy themselves in a war, and if they need to regrow a limb or change themselves in some other way they'll be able to do it. And there will be other benefits.'
>
> 'But they won't be human,' Lilith said. 'That's what matters. You can't understand, but that's what matters.'
>
> Its tentacles knotted. 'The child inside you matters.'[23]

Dawn could be read in the sense of hierarchy and control over other species hidden behind the idea of a savior. Throughout the story, there are many mentions of animals, specifically laboratory animals who are kept in a cage, messed with and controlled to create knowledge about a specific physiological, metabolic or genetic process. And, also as Lilith believes they are part of an endangered species in need of protection in the midst of a conservation program, suggesting a process of saving a species from extinction, by extracting them from their land and messing with their genetic components to the point of hybridization.

The parallel can be exemplified by the paradigmatic case of the conservation of Northern White Rhinos. The Northern White Rhinos' story of conservation has had many ups and downs, throughout the 1960s and 1980s, they were hunted to near extinction by civil war dynamics and poaching practices for illegal trade of animals and animal parts (especially their horns). During the 1990s, many conservation areas were created in Kenia. In particular, Ol Pejeta Conservancy took a longer time to be developed, but by the beginning of the 2000's this rhinos' population had increased with their protection.[24]

Sadly, the poaching did not end, hence in 2006, only six northern white rhinos remained. Two of them were relocated to San Diego Zoo and soon found to be not eligible for reproduction conservation. The other four, after some time in the Czech Republic, were transferred back to Ol Pejeta Conservancy, where all remaining males died in 2014 and 2018, leaving only two females: Najin and Fatu. Who have been intervened multiple times for assisted reproductive technologies. Since then, many assisted reproductive technologies processes have been tested: ovum-pick up, intracytoplasmic sperm injection, embryo transfer and, as of lately (2024), in vitro gametogenesis from induced pluripotent stem cells derived from reprogrammed fibroblasts.[25] The last one points to a hybridization process with Southern White Rhinos.

There are many ethical challenges to be explored in these conservation practices based on assisted reproductive techniques. In the article *Ethics at the edge of extinction*, Pierfrancesco Biasetti, Thomas B. Hildebrandt, Frank Göritz, Susanne Holtzel, Jan Stejskal, Cesare Galli, Daniel Čižmàr, Raffaella Simone, Steven Seet and Barbara de Mori Biasetti, consider an ethical matrix to evaluate three main elements in conservation efforts: biodiversity, the animals and the people. Through the elaboration of the matrix, they point to several factors at play in each element. I would like to highlight three of them:

First, the critiques of conservation efforts with reproduction technologies have shown these interventions do not attend the real issue underlaying extinction but only show a short-term technofix. Second, the genetic framework to establish the value of a species (or in this case sub-species) as biodiversity in which a hybrid may not be "as valuable as a pure one." And lastly, these efforts are not always focused on the individual animal's well-being since they are subjected to many medical procedures (although they also highlight the benefits of being under veterinary care in general). In this sense, I would like to add to these issues an important idea: the imperative of having a life worth living.

Regarding the first issue, and following Taylor's stories of disability, it is fundamental to address the root causes of disability in an ecosystem or an animal population. In this case, poaching and the illegal traffic of animals, even if supremely difficult to stop entirely, should be a substantial concern of conservation practices and mitigation can not only be focused on asking people to stop the demand. Of course, this problem is a huge one to tackle, and the point is not to criticize the work of people in natural conservation reserves, they do amazing jobs and, in my view, they deserve recognition.

What I do believe must happen is that conservation biologists, practitioners and theorists, could address hunting and illegal trafficking more directly than they usually do. Barbara Noske, has been critical of the lack of political stance against hunting in the conservation and environmental community, something that is mostly thought of by the animal side of academia (including anti-speciesism and animal liberation scholars).

This lack of stance may be a factor in perpetuating the pessimistic and "nothing to do" attitude towards a highly difficult problem to address. Yes, there is probably not much we can do to immediately stop poaching and illegal traffic of animals, but we can expose: "how systems of power rely on, benefit from, and produce disablement in and among human people in ways that are enmeshed with how they benefit from and produce disablement in and among ecosystems and non-human species."[26]

The second and third concerns might be connected, the idea of genetic conservation focus on a biodiversity made of valuable genes that should be preserved, takes the interest away from the individual animals and their well-being. What is valuable of being conserved, genetic biodiversity, does not necessarily imply individuals. Then, the practices with which animals are intervened by reproductive technologies are seen as necessary for the preservation of a perceived valuable set of genes. Still, many questions remain: Who benefits in these interactions? Are rhinos interested in the extinction of their taxon? If we could ask, would they be willing to submit themselves to all the medical procedures done to them throughout their lives, in order to preserve biodiversity? Would we?

Considering these questions in the light of the Oankali endeavour of saving the wildlife of Earth, it is clear there is an important form of biopower being exerted by the Oankali in the profound sense of not only modifying humans' genome but, also, renewing their world and modifying many other species. Specifically, the modification of their reproductive capacities is a biopower practice that connects the Oankali with other conservation programs focused on reproductive technologies. The Oankali though, are not hierarchical beings, meaning any action they take is not based on a subject–object relationship, and it is not epistemologically tied to biology and biotechnology in the human sense. Also, Oankali were not there to keep everything as it was, they are beings of combination, diversity and change, in a sense they are the ultimate genome captors and re-combinators.

Conservation practices, on their behalf, are in some ways trying to keep and maintain some sort of past or present environment and beings. Conservation, as the name suggests, is not rallying for change but for a sense of stability and equilibrium. This also does not mean that wildlife in conservation projects is not slightly modified to adapt to the new and rapidly changing environment. Many conservation projects involve reproductive technologies that were born to attend and use animals' reproductive capacities in several industries. Animals are being modified by these practices, their reproduction capacities are being used, for their supposedly own good, or the populations and environments'.

Maybe as we humans see it, being on the exerting end of the stick, wildlife conservation efforts are grand practices of searching for wellness for other animals and their environments, and in many ways, I do believe they can be. But also, we must admit they involve a certain arrangement of understanding ourselves as not only able but compelled to mess with animals' lives, to protect them. As extinction grows, conservation grows with it to repair the damage caused. But is there an acknowledgment of the root causes? And when there is, who is affected and who is to be accountable for?

Lilith and all the other humans felt they were being robbed of their humanity, in the sense of their autonomy, free will, control over their bodies, movements, feeding practices and environments. Even if some of them understood the actions of the Oankali as something close to a savior, they still did not like being changed, touched and paired without their consent, or being sterilized and impregnated without their knowledge.

The idea, again, is not to bash conservation practices, but to critically engage in the dilemmas they face. One way of thinking of the effects these practices might have is to think we are not on the exerting but on the receiving end. This, on an optimistic view, would help strengthen the effectiveness of conservation efforts instead of simply appealing to stop them. A disabled ecologies framework may help us question and fortify the understanding of potential treatments, access and care as a question of how we would like to be treated[27] and, in this sense, to apply how we would like to be treated to other beings.

2. On the Place of Disabled and Polluted Animals, on *Barn 8*

> Where would they take these birds where they'd be free? The chickens were so overbred, they no longer had a natural habitat. 'You have to have a place to go where you can be free, in order to be freed'.
>
> —Deb Olin Unferth, *Barn 8*

Barn 8, by Deb Olin Unferth, is the story of a mass robbery of chickens gone wrong, or at least wrong for some of the implicated beings. I will not delve into the whole story; I will only focus on the ending and the epilogue.

The chickens in question do not live on a ship in between planets, they have their own machine universe, described as:

> Inside: Barn Universe. Completely enclosed in steel and concrete, seven tremendous aisles of cages soaring twenty-five feet high, eight tiers in two stories. A system of chains bringing in the feed, a series of belts carrying out the excrement. Powerful fans pulling through and out the carbon monoxide, the hydrogen sulfide, the ammonia, the dust. Twenty thousand one-foot candle bulbs at regular intervals, like a monstrous Christmas decoration, the sun rising and falling on a timer. The entire barn rumbling with machinery. A hundred and fifty thousand chickens stood there, waiting – for what? For whom? On a wide conveyor, eggs slowly floated by.[28]

The description of Barn Universe points to the huge machinery involved in farming industry nowadays. In this sense, this is not a fantasy world or a ship, but the actual place and universe chickens have been assigned to for many generations. It reminds us of how humans have selected and expanded a place for these animals to live in, so that they can be used to produce eggs and meat. So, it points to an important issue regarding disabled ecologies stories, they shine light on, not only who have been injured and how (being kept and managed inside the Barn Universe), but also "that they are disabled by the ways society is organized – by a system of economic growth."[29]

These animals, as well as many others, are trapped by the Industrial Animal Complex (IAC),[30] they are property and stock numbers. They are consistently and productively disabled by this Barn Universe, by the places they live in and what they are fed with, immunized from, what they breathe and their severely modified reproductive capacities. They are, in a sense, not so different from the humans in *Dawn*. Only they have been trapped for so many generations, they do not know anything different. When they are liberated, when they are invited to land they are confused:

> No one had ever asked them to come have a look at the sky, and they couldn't see it from their cages. They'd never felt space around them, never dirt or grass under their feet. The hens paused just inside the crowded barn and gazed out, heads tilting and turning. [...] The air was moist and clear in ways unfamiliar to them- Pollen floated through it. That's the thing: the want the air – cool, bug-filled, quiet – and they will soon love the rest, if you just give them a little more time. Let them adjust.[31]

Living all their lives inside the Barn Universe, they have lost any possibility of being "sent back" to nature, as there is no natural ecosystem in which they belong or that can sustain them. The chickens were freed on their own in a polluted piece of land comprehending an old barn now unused, next to a chemically contaminated Natural Reserve in which no human should enter and is only managed by one park ranger. When the park ranger sees the chickens roaming around the reserve she is puzzled. First, she is amazed that so many of them were still alive under the polluted landscape, upon reflection she understands:

> [...] the little ladies are underrated. These hens have been bred through so many generations -a sort of sped-up evolution – that they've become genetically strange and powerful animals, designed to live through rather bizarre trials. [...] They have to be able to just stand there in those cages for months and months, their lungs filling with dander. They have to be able to take the onslaught of massive doses of inoculations, to endure sensory overloads and deprivations, clown-car crowding, the vicious pecks of their cell mates. They have to be resistant to disease. They have to tolerate violence, noise, panic, and no drop dead of heart stress (as many do, apparently). These birds are practically radioactive, if you understand what I mean.[32]

Curious about their possibilities, and believing they are extremely rough animals, she decides to buy roosters to give the little ladies a chance to survive in a land that was not going to be used any other way. She believed their chances were slim, but hoped their primitive genes were strong enough "to conquer all that humans had done."[33] So that they would want to lay eggs and care for them, something that was stripped from them many generations ago, and to "be moved enough to break into the ancient twittering song that hens sing to their embryos, songs that the embryos sing back."[34] And, they did. As we learn from the epilogue: "The chickens outlived the humans."[35]

The chickens then liberated and paired with roosters, survived and thrived for 50,000 years. They did not only survive in the polluted reserve, but they also became highly good at storytelling. After all, chickens today and outside the Barn Universe may have more than thirty topics of conversation and complex social lives. Inside, their ancestors "passed down only the imaginary and sensation of claustrophobia, rotten air, debilitating hopelessness, and pain, so that sadness lived within their tribe."[36]

Once freed, they developed storytelling capacities, so much so that they made an annual pilgrimage back to the barn their ancestors came from, to their origins: "As turtles crawled back up the sand, children come home after they go out to seek their fortunes and fail, creatures return before they start stuck out for more, so the chickens, to mark the land with their eyes. Instinctual, biological, psychological, spiritual – anything but intellectual – is what we living things do: look back."[37]

The novel ends with this hopeful phrase: "They'll take only what they need. They'll run over the land, eating the grasses and the bugs that survived and revived and thrived. They'll live."[38] The path of transformation sketched in this epilogue is an interesting answer to the question: what if chickens were left to be free. Many paths could be taken from there but the fact that chickens are able to survive in such a contaminated place because of the multiple disabling interventions they have experienced for generations (and still experience to this day) in the Barn Universe, makes me feel uneasy.

Framing it with the stories of disabling ecologies, there could be a reflection left thinking this ending with the idea that: "[i]n this age of Disability we can fear disability and see it as the antithesis of the future we wish to build, or we can center the damaged, the vulnerable, the ill and the injured, the contaminated and the poisoned, and build a new world from there."[39] In *Barn 8*, the vulnerable (ecosystem and animals) were left to build a new world without humans continuing to contaminate them, and they thrive.

However, centering the vulnerable means to actively engage with healing, supporting and caring activities, not only by letting them be free and hoping that maybe one disablement and other will pair up perfectly in order for them to survive together. In *Barn 8*, there is no mention of how and if the environment, the reserve, changed with the chickens flourishing, but we can assume that it did, given that it was stable enough to keep the chickens alive for 50,000 years. Yet the park ranger understands these chickens as practically radioactive, strong and adaptable, not as disabled and vulnerable animals, who are not likely to survive if they are liberated in an equally polluted or disabled land or territory; and this belief is what makes her intervene.

This circumstance points to difficult questions like: How should we intervene if we must? Should we humans just take a step back and give animals land to thrive in? What can we do if we do not disarticulate the Industrial Animal Complex and all the background it comes with? How do we treat the injured? How do we want to be treated if injured?

Barn 8 epilogue has always reminded me of one particular and popular case of animals surviving in a highly polluted area: the dogs of Chernobyl. A disabling story with many tangents and deeply engrained in many people's lives. We are probably all aware of the nuclear catastrophe in Chernobyl in April 1986, the massive explosion burned for ten days releasing several radioactive chemicals across Ukraine, Belarus, Russia and North America. Today the area surrounding the nuclear power plant is now known as the Chernobyl Exclusion Zone, which has been labelled and managed as a Natural Reserve.

When the explosion occurred, people on the ground were directed to leave everything behind, including any pets they cared for at the time. Once people were evacuated, the dogs were meant to be annihilated out of fear of what the radiation could do to them, and if they could potentially spread radiation outside the exclusion zone. But many of them survived and have since lived in the area, creating several groups and even accompanying people who work on the checkpoints of the area.

There are two reflections I would like to make whilst thinking about *Barn 8* and the dogs in Chernobyl. First, the case in which even in a heavily polluted, radioactive area, many animals are able to keep on living and even thriving when the area is considered a Natural Reserve in which human activities are no longer managing the land for their own purposes. Giving animals land to live in, giving them space, even if it is not the best land, gives them a chance to survive on their own. This is clear in both stories.

I am curious though, if this narrative has underlying discussions. A question that arises in both stories is why a contaminated area is declared a Natural Reserve. Of course, it is a way of keeping humans away from polluted areas for their own health, but it also creates a restriction for construction and industrial growth in that land; creating possibilities for other beings to flourish because, even if polluted, the land allows them to continue living and growing; something that is not possible if an area is not restricted to human activities.

Also, because these preserved areas can be fundamental for experimentation, they create isolated landscapes of chemosociality, a term brought forth by Eben Kirksley, that refers to the relationship between beings and places that have been "shaped by chemical weapons industries, municipal landfills, government remediation programs, real estate speculation and a multitude of chemical and biological agents."[40] Chemosociality points towards the alteration of relationships emerging in these chemical-exposed lands and beings. In the case of the dogs of Chernobyl and other fauna and flora living in the Exclusion Zone, they are continuously monitored to understand the effects of radiation, the mutations occurring and the possibilities that humans have to reclaim space.

Unlike the chickens in *Barn 8*, Chernobyl dogs are not entirely away from humans, they live among them, they have bonding relations with people on the ground. An example of which is Ludmilla, a person who starred in the documentary film *Fallout dogs* by Julia Oldham (2019), and many of the guards in the check points of the exclusion zone. People and dogs' relationships, as Jonathon Turnbull identifies, are enmeshed in a complex matrix of meaning and accountability; it is a symbiotic relationship working in between detachment and engagement, in a care-full relationship.[41] Humans generally offer them food, and dogs provide protection, company and entertainment to the guards. In this sense, the author reminds us that:

> Paying attention to the everyday lives of humans and animals together in toxic landscapes allows for the nature of these spaces, and the animals that dwell there, to be rethought beyond the filthy vs flourishing binary.[42]

These relationships can be also thought in the sense of chemosociality, they are reconfigurations of relationships on polluted, toxic and disabled ecosystems. The idea that just by leaving animals free and stopping human intervention is not a black and white situation, there are moments and places where human intervention is not needed and others where intervention is possible and may give way to a relationship in which we can all learn to live with each other better.

Both stories show that the intervention that matters is the fact that these areas are isolated and unproductive for human industrial activities. Once turned into natural reserves, once human activities are not allowed, animals and ecosystems can rebuild their environments, with or without the help of humans. Nevertheless, only allowing animals and whole ecosystems to live on polluted land, while the rest of the land is made for and owned by humans, is a root problem to address if we are to think of how to mitigate, support, care for and heal disabled ecosystems and communities.

Even if in both stories, chickens and dogs survive on highly polluted spaces, we must remember "[o]nly some species are able to persist in conditions of altered life, with life altered by chemical exposures and histories of colonialism, warfare, and capitalism."[43] Then, we must not only think that some organisms might survive global extinction processes because they will adapt to an altered and disabled environment, but that all organisms could be in the umbrella right to be rescued, as Taylor proposes.

As I have tried to convey in this chapter, extinction processes are complex and multifaceted. Extinction caused directly or indirectly by humans' activities on the world we live in, are the ones I focus on the most, because following Taylor's disabled ecologies framework we are to pay attention to the ways in which we are responsible for the disabling of land and whole ecosystems. We should mind the root causes and make industries accountable for the destruction they create. In doing so, we must engage in critical dialogue with those who work in conservation and rewilding practices, to find better ways of renewal, healing, mitigating, and supporting. What we do matters, what and who we value matters, how we attend this crisis is fundamental for the ongoing of life on Earth. We must keep thinking about extinction processes, not only to show the damage done, but to imagine ways to remediate and heal. As Haraway suggests:

> What happens when a partner involved critically in the life of another disappears from the Earth? What happens when holobionts break apart? What happens when entire holobiomes crumble into rubble of broken symbionts? This kind of question has to be asked in the urgencies of the Anthropocene and Capitalocene if we are to nurture arts for living on a damaged planet.[44]

Notes

1 Depending on the ecofeminist's author standpoint the term 'women' is not necessarily defined and limited by biology, hence it is not an essentialist concept. From the standpoint of this book, this devaluation of women is understood, from a queer and intersectional position, as a bigger more inclusive system of power named cis-hetero-patriarchy. Or the system in charge to devalue everyone who does not fit the cis and heterosexual white male mold.

2 Donna Haraway, "Situated Knowledges: The Science Question in Feminism and the Privilege of Partial Perspective", Feminist Studies 14, no. 3 (1988): 592.

3 Rewilding is the name given to a process in conservation of nature in which the goal is to restore an ecosystem by different means like reintroducing species, protecting habitats and reducing human impact.

4 Rosi Braidotti, *The Posthuman* (Cambridge: Polity Press, 2013), 164.

5 Donna Haraway, *When Species Meet* (Minneapolis: University of Minnesota Press, 2008), 18.

6 Sunaura Taylor works with critical disability studies, in this sense, the concept of disability she uses might require a little context for people outside the field. Disability is a multifactorial concept, and it is not seen in these types of studies as a moralizing concept per se (good or bad). This does not mean that people with disabilities (neuro-developmental or physical) don't suffer from the conditions they live with (like chronic pain, restricted movements, anxiety and many others). It means there should be accommodations for people with disabilities to experience the world more fully, we should try to find ways to heal and support them. It does not mean, though, that a person with disabilities can be considered wrong or less than someone without disabilities. Taylor also uses disability closely to the "social disability" framework created by different analysis in diverse fields and frameworks of social sciences (public policies, sociology, psychology, administration, and others; with Marxist, Foucauldian, feminist, queer, and other frameworks). In general, what they highlight is that disability is a product of a social and material organization of the world. If

interested I would recommend reading either of Sunaura Taylor's books, *Beasts of Burden* or *Disabled ecologies.*

7 Elizabeth Kolbert proposes that actual human beings face a Sixth Extinction caused by anthropogenic (human made) activities.

8 Octavia E. Butler, *Dawn* (London: Headline Publishing Books, 2022), 43.

9 Ibid., 40.

10 Ibid., 36.

11 Ibid., 34.

12 Sunaura Taylor, *Disabled Ecologies: Lessons from a Wounded Desert* (California: University of California Press, 2024), 7–8.

13 Butler, *Dawn*, 41.

14 Ibid., 211.

15 Ibid., 45.

16 Ibid., 67.

17 Ibid., 109.

18 Ibid., 157.

19 Ibid., 201.

20 Ibid., 218.

21 Ibid., 223.

22 Ibid., 280.

23 Ibid., 282.

24 Pierfrancesco Biasetti, Thomas B. Hildebrandt, Frank Göritz, Susanne Holtzel, Jan Stejskal, Cesare Galli, Daniel Čižmàr, Raffaella Simone, Steven Seet, Barbara de Mori, "Ethics at the Edge of Extinction: Assisted Reproductive Technologies (ART) in the Conservation of the Northern White Rhino", Journal of Agricultural and Environmental Ethics 38, no. 7, (2025): 3–5.

25 Ibid., 5.

26 Taylor. *Disabled ecologies*, 29.

27 Ibid., 242.

28 Deb Olin Unferth, *Barn 8* (Minnesota: Graywolf Press, 2020), 59.

29 Taylor, *Disabled ecologies*, 114.

30 The Industrial Animal Complex was proposed by Barbara Noske as an umbrella term that recognizes the variety of animal industries that exist nowadays: meat and derivate animal products, leather and fur industry, medical and beauty industry, education programs and biotechnology industry, as well as several entertainment possibilities (circus, zoos, pedigree competitions, etc.). This term also allows her to explain the integration and interconnection of the animal industries with capitalism, whether historically, locally or globally. See book: Barbara Noske, *Human and other animals* (London: Pluto Press), 1989.

31 Unferth, *Barn 8,* 248.

32 Ibid., 263.

33 Ibid., 270.

34 Ibid., 270–271.

35 Ibid., 281.

36 Ibid., 282.

37 Ibid., 281.

38 Ibid., 282.

39 Taylor, *Disabled Ecologies*, 278.

40 Eben Kirksey, "Chemosociality in Multispecies Worlds: Endangered Flogs and Toxic Possibilities in Sydney", *Environmental Humanities* 12, no. 1 (2020): 23.
41 Jonathon Turnbull, "Checkpoint dogs Photovoicing canine companionship in the Chernobyl Exclusion Zone.", *Anthropology Today* 36, no. 6, (2020): 23.
42 Ibid., 24.
43 Kirksey, "Chemosociality…", 32.
44 Donna Haraway, *Staying with the trouble: Making kin in the Chthulucene* (Durham: Duke University Press, 2016), 69.

Chapter 4

EXTINCTION: FROM BIOPOWER TO BIOPOETICS OF LIFE

In *Extinction* documentary series, there is a chapter dedicated to telling the story of the Tasmanian Tiger (Thylacine), specifically the story of its extinction. The chapter opens with a scene in which one man equipped with a shotgun kills what could have been the last specimen of the Tasmanian Tiger. The narrator's voice, after the shooting, precises that the killing cannot be reduced to ending the life of a single animal, because "it marked the death knell for an entire species."[1] This is one piece of evidence that extinction events are driven not only by natural process, but by human activity. The scene shows that with the death of one single specimen an entire species could disappear from the face of the Earth. Besides, the killing of the tiger is performed with the force of a technical object. This act displays on a small scale the terror of using this type of technology. This act occurred in the 1930s and could be a paradigmatic example of catastrophic events that would lead to the extinction of other species or at least place them in danger of extinction using other technical objects.

However, extinction of species may take other paths. Take as an example what Menno Schilthuizen has studied around what is called Roadkill. Schilthuizen ironically refers to this act as *roadkill literally "drives" some species to extinction.* And this is a case in which a technological object, such as a car, used for transportation, may present a risk of killing animals of different species on the road. To make it more dramatic, think of all the busy roads around the world, where the death of thousands of animals occurs daily. Probably, this act is not direct, because humans do not seek to kill animals; and even ignore that they ran over a specimen that might be in danger of extinction. But this does not take away from the fact that humans using cars drive certain species to extinction.

These apparently innocuous two examples show that certain technological objects play an important role in the current process of extinction. Besides, they expose that humans whether aware or not, directly or indirectly, in slow and even piecemeal processes, can drive some species to extinction.

Moreover, humans have become a geological force that is changing Earth's climate, by which many species, including human species, are or will be in danger of extinction and probably become extinct in the near future as critical planetary dynamics are changing. This current process of extinction has been recognized as the sixth mass species extinction in planetary history.

It is worth asking the question: can this contemporary process of extinction be analysed within the biopolitical frame? In a sense, the answer would have to be negative, because biopolitics has rather been tasked with strengthening life and increasing its forces, while administering and managing it. However, does biopower only exert power to foster life or make live? Evidently, the killing of animals either by shooting or being run over by cars are not effects of a form of power of fostering life or making live. Neither are they a form of power to let die. If the current process of extinction has the human mark or show the human footprint, then it is possible to understand that humans are throwing not only into death but also to extinction of some specimens and species. The evidence of this form of power can be observed by the biohistorical records that list every threatened or extinct species, just as the International Union for Conservation of Nature and Natural Resources (IUCN) Red List. Now that humanity is facing a series of species extinction caused by anthropogenic activities linked to global warming, can the situation be changed? Or is humanity condemned to see how different species become extinct? Is it possible to change the biohistorical record of extinction? How to create another perspective that resists biopolitics and transforms biohistorical interference between life and history? In front of biopolitics and biohistory, is it possible to put a form of biopoetics, understood as the way by which interference between life and history would happen in the future? To try answering these questions, the chapter is divided into the following sections: the first section entitled *A glimpse of (de-)extinction*, based on the novel case of the de-extinction of the Dire wolf, shows that biopower makes live, lets die, throws to death and extinction; the second section entitled *The biohistorical record of extinction* considers, in biohistorical terms, the importance of lists and records of extinct species, as well as the problems they face that render them inoperative to some extent, to demonstrate the ways in which the movements of life and the processes of history interfere with each other. In the final section, entitled *From Biohistory to Biopoetics: Avoiding Death and Extinction*, working with the interference between life and history, reviews the gallery of biopoetic perspectives from different theoretical fields to propose a way in which the processes of life could become a historical movement to avoid extinction.

1. A Glimpse of (De-)Extinction

Since Dolly, the first cloned sheep, human beings have realized that the production of life only needs a few materials: DNA, a technological set known as laboratory and a group of scientists who have the grace to do science with life. Through these materials, life can reproduce itself or, in other words, life reproduces itself by other (technical) means. Is it possible to take off the human agency in this kind of practice to put the agency in life itself? In this case, are humans instruments for reproducing life itself or is life a human instrument? Perhaps, in modern times, when humans are the only subject aware of themselves and of the world of life, it is impossible to think of life's agency, even though life is what has produced the evolutionary line by which humans have emerged, in the recent historical period (say, two and a half million years). To give some credit, if life can be reproduced by human hands (techno-hands), it is because life can transcend itself since the beginning. In Simmel's metaphysical perspective, Dolly is not only life that is more-life but more-than-life because it is a spiritual product that cannot be reduced to organic material. With technological support, life embodied in Dolly has transcended the natural boundary. Certainly, the spiritual realm does not only place all the necessary agencies into producing more-than-life but gives humans the responsibility to be in charge of production. So, when humans transcend the more-life dimension, it is easy to admit that life, natural life in this case, lacks agency, because it cannot be responsible for acting. Life just acts without a sense of morality.

After Dolly, every cloning practice in humans, at least, was forbidden because of that principle. But so far in the twenty-first century, now it is possible to clone dogs, cats, horses, camels, etc. That is to say to produce another form of life. Now, the exercise of power over life produces both bodies and lives through genetic technologies. It is easy to understand that domestication and, in this case, cloning are both effects of biopower. Even cloning could be the new expression of traditional domestication guided by the principle of (artificial) selection.

Recently, this exercise of biopower has produced another effect that cannot be limited to the reproduction of living forms of life. Just like the sf story *Jurassic Park*, recently human society witnessed how a biotech company (Colossal Biosciences) has brought to life the Dire wolf (*Aenocyon dirus*), which had been extinct for more than 10,000 years. In fact, the three wolf specimens brought to life are hybrid species, not only because of the usage of the Gray Wolf's DNA, the most recent relative, because of the creation of this life form through cloning technologies and genetic engineering.

Considering the Dire Wolfe as the first example of de-extinct animal, it can be realized both the recent effect of biopower and the projection into the future of being able to bring extinct species back to life. So, the twenty-first century perhaps will be acknowledged as the century of multiple de-extinctions. Even more so, because humans, being blamed for causing the extinction of many species, will acquire responsibility for resurrecting and reintroducing them into their respective habitats (the invocation of spiritual realm that transcends the natural boundaries, just as the individual life span and the species' life expectancy). In other ways, humans are becoming responsible, as never before, for life.

This principle of responsibility leads to another important discussion, a retrospective one. If throughout its history life has produced many life forms that eventually become extinct, nobody blames nature for establishing this normal dynamic. For sure, the extinction of species is the deathliest natural process in life's history. At the same time as life produces life forms, it also creates extinction latency in those forms. No matter if for thousands of years one species manages to survive successfully, at some unexpected moment, it will tend to disappear from the face of the Earth, leaving only, at best, its debris: fossils for specialized eyes.

So, life forms reach their existential limit when they become extinct. That's the moment when life transcends itself and turns into death. This transcendence of life itself, understood as a natural process, goes hand to hand with death and extinction. In fact, in present day, Earth's history all the life forms that have become extinct might be counted. It is estimated that, of the total 100% of species that existed, only 4% are currently alive. Out of that 4%, around 8.7 million species, more than 47,000 are threatened with extinction in recent years.[2] But now, thanks to the new field of research that has applied the concept of Anthropocene that points out that human beings are the main problem related to extinction issues, it is not only about the historical footprint that has marked nature with culture, but that in some sense human culture has specialized in extinguishing life forms either directly (as the Thylacine) or indirectly (as the dodo). Because of this, extinction as a natural process has also become a cultural one. Human activities, centered in Western, capitalistic and hyper technical culture, produce anthropogenic effects that have affected each stratum of the Earth, such as soil, water, climate and air, and subsequently to all life forms. In the Anthropocene, it is not that the natural process of extinction has been replaced or surpassed by cultural processes but only is produced by other means in which the exercise of power takes a leading role.

The Dire wolf case has shown that the power over life produces animal bodies from the use of different DNA to de-extinct only one species. But does this mode of power only produce life, or can it throw it into extinction? This new De-extinction effect is the novel form of biopower that seeks to produce life through biotechnologies. One effect of this mode of power is to produce bodies and to control the biological features and process of the living entities and another one that is not only complementary but extensive, corresponds with the possibility, in appearance, of bringing back specimens of some extinct species. These effects of power show one of Foucault's main theses, that power is not repressive but productive. In this case, biopower, which began in the nineteenth century, is productive in the sense that it produces life. Under this consideration, the Dire wolf case shows the productive effect of biopower. It does not matter if it is a real de-extinction[3] or only a mix of DNA, belonging to different species, that has created hybrid animals, the very important situation in the biotech practice is that it is effectively producing life under controlled conditions. This case is not only a technical-scientific success but a result of a historical exercise of power and more precisely an exert of power making life live.

Now, it is important to ask whether this exercise of power not only *produces* life – in the sense of "making live" – but also exerts a form of power that, on one hand, *lets die*, and on the other, *throws* something living into death. This distinction between *letting* and *throwing* is not immediately clear in Foucault's famous formulation, especially considering that the standard English translation from the French emphasizes the former action: *to make*. This translational choice may reflect a binary perspective that foregrounds active versus passive positions in the exercise of power. In such a view, *making live* represents an active posture of power, while the passive posture – characterized by non-intervention or restraint – is seen as unproductive. In this framework, unlike sovereign power, which "lets live and makes die," biopower – closely tied to specific forms of knowledge, particularly biological knowledge – is understood as productive insofar as it entails an active *making live*.

However, this productive aspect of biopower can also lead to death. Foucault employs the French verb *rejeter dans* to describe the second part of his formulation, a phrase that aligns more closely with the English verb *to throw*. Hence, *rejeter dans la mort* may be more accurately rendered as "to throw into death." Accordingly, biopower emerges as a form of power that either *makes live* or *throws into death*. If *letting die* carries a passive connotation, *throwing into death* implies an unmistakably active stance. Thus, the subject who exercises biopower occupies an active position – whether by fostering life or by forcefully expelling the living into death.

Once the era of biopower has begun, the power of making live or throwing to death exerted over two types of bodies. One related to the mechanical behaviour of a body considered as a machine produced by disciplines and the other one related to the mechanisms of the living body considered as body-species which are produced by regulative controls. The first is focused on individual bodies and the second on the body of a population. So, according to Foucault, biopower could make live both types of bodies while could throw to death other individuals or populations.

Characterizing biopower as a form of power that could throw to death individuals or populations is very important in extinction issues, because it can drive the analysis in a deeper sense. Maybe Foucault did not realize the full scope of its formulation, because when he characterized the body-species, he thought only in terms of population and not in the biological concept of species. In some sense, he coined the concept of body-species only to claim that biopower is not reduced in controlling the body's anatomical features, but in controlling the biological process too. In this sense, it does not mean that biopower exerts on the whole species.[4]

But now, thanks to the statistical records about extinction of species, it is possible to claim that biopower effectively exerts on the biological level of species. It can scale from individual specimens to populations and then to species. If this is the case, then the power of throwing to death (individuals and populations) becomes a power of throwing to extinction (species). In the present mass extinction, humans are exerting, in one hand, a power of letting die because they do not stop affecting the planet with their polluting excrescences which indirectly are killing several animals and, in other hand, a power of throwing to extinction, because despite knowing of the threat of extinction, they continue to kill living beings to the point of throwing them into extinction. There is no better example of the form of power of letting die than the one used by Michael Hoffmann for referring to the function of Alliance for Zero Extinction (AZE) that aims to conserve places and key sites where endangered or critical endangered species reside: "These are the places where, if you don't do something here, for this species, you're going to lose a species."[5] And to understand the exercise of power that throws into extinction, take as an example the Mexican wolf's case that in a certain time was extinct in the wild due to intensive hunting by humans to protect their livestock. If it were not for rehabilitation practices in captivity, that species would have become extinct.

2. The Biohistorical Record of Extinction

> Once on the brink of extinction, the individual becomes larger than a singular life, and her or his death is therefore more than a singular death – it becomes the death of a life form, the death of nature.
>
> —Irus Braverman, The Regulatory Life of Threatened Species Lists

In 1974, in his lecture titled *The Crisis of Medicine or the Crisis of Antimedicine?* given at the Institute of Social Medicine, Biomedical Center, of the State University of Rio de Janeiro, Brazil, Foucault stated that human beings have entered in a biohistorical time thanks to the progress in the field of genetic engineering. However, entering into biohistory is not only due to techno-scientifical progress, but by acquiring the ability to modify the human species at the genetic level in which techno-sciences play a main role. Taking as an example the history of medicine, Foucault explained that traditional medicine produced certain effects over the individual and in a wide spectrum over the family and its descendance; but modern medicine related to molecular biology and genetics could affect both human and life itself:

> Nowadays, with the techniques at the disposal of medicine, the possibility for modifying the genetic cell structure not only affects the individual or his descendants but the entire human race. Every aspect of life now becomes the subject of medical intervention. We do not know yet whether man is capable of fabricating a living being which will make it possible to modify the entire history of life and the future of life.[6]

Foucault called this historical process biohistory, because medical specialists and biologists "are beginning to work at the level of life itself and its fundamental events."[7] But considering the Darwinian Theory, Foucault claimed that human society across the time could modify flora and fauna of certain spaces. In consequence, the history of human beings and life are "profoundly intertwined."[8] Just as Foucault said: "The history of man does not simply continue life, nor is simply content to reproduce it, but to a certain extent renews it, and can exercise a certain number of fundamental effects on its process."[9]

From the perspective that Foucault offers, it becomes easier to grasp how the most recent effects of power (such as de-extinction through cloning, hybridization and genetic engineering) represent a fundamental modification in the history of life. These are not mere reproductions of life, but rather renewals of it (assuming the specimens survive). Within the framework of biohistory, the exercise of biopower consists in producing life by *making live*. From Dolly the sheep to the recently recreated specimens of the Dire wolf, one could argue that contemporary humanity is now, as Foucault suggested, operating at the level of life itself and producing effects at its most fundamental strata.

To understand another aspect of biohistory, these two distinctions and precisions could help: (1) If for Foucault biohistory is constrained to studying "the effect of medical intervention at the biological level, the imprint left on human history [...] by strong medical intervention that began in the eighteenth century,"[10] then it is time to study the effect of human history or human species history intervention at the biological level, the imprint left on life itself by the strong medical, biological, genetic and biotechnological interventions that have occurred since the second half of the twentieth century. (2) If for Foucault biohistory has to be interested in "the evolution of relations between humanity [l'espece humaine], the bacillary or viral field, and the interventions of hygiene, medicine and the different therapeutic techniques,"[11] then the main goal of biohistory in the twenty-first century is to study the evolution of relations between human species, other species and the interventions of medicine, biology, genetics and biotechnology.

In consequence, using Foucault's toolbox but transcending the human-centric frame, because human species is not the only one to suffer the effects of power and biopower, a biohistory for the twenty-first century must study the three dimensions of the exercise of power over life: making live, letting die and throwing to death and into extinction. This form of biohistory must consider that humans belong to a living species, among others, and inhabit a living world.[12] In doing so, the question arises: At what price do human beings make a species live, let it die, or throw it into extinction?[13]

In this case, it is easy to trace the biohistory of the power of *making live* when it is related to the production of life and bodies of other species or animals. But in the case of the mass extinction driven by the anthropogenic footprint, how is it possible to trace the biohistory of the power of throwing to extinction? It is important because it can show that human history not only continues, reproduces or renews life or certain species, but exercises a certain number of fundamental effects on its processes even to the point of interrupting them forever, either directly or indirectly. If the eighteenth century, just as Foucault asserted, is the historical moment when life entered history, then the twentieth century is the historical moment when the extinction of life (as produced by humans) entered into biohistory.

Two centuries since the beginning of the biopower age, during the Second World War, the use of the atomic bomb showed that life is easy to suppress, as Foucault points out: "So the power that is being exercised in this atomic power is exercised in such a way that it is capable of suppressing life itself."[14] But in this case, within the human-centric frame, life itself is related only to human life or, more specifically, life of one human population, even if the effect of destruction suppress nonhuman life (plants or animals).[15] Transcending this humancentric frame, this type of power could not only destroy the life of a certain population of animals or plants but suppress life itself. Consider the nightmare produced by atomic power, during the Cold War, that implies the destruction of the planet and consequently of life itself. Even today, the threat remains latent in armed conflicts between countries that produce nuclear weapons and in other fields such as accidents at nuclear power plants as those that occurred in Chernobyl (1986) and Fukushima (2011), whose extended negative effects to the environment affected lives beyond the human one. In this case, to suppress the whole life it needs a huge explosion, whether due to global war or due to carelessness at a nuclear power plant that triggers a catastrophic global event. This event could probably kill life itself.

However, in the twenty-first century, extinction, so far, is not occurring by genetic engineering nor atomic power, but by the wide transformations global warming is producing. This planetary event, according to the Anthropocene theorists, is producing a massive extinction of species and in the long term, will produce the extinction of life itself. What was started by humans will be finished by temperature changes that will make the planet an unlivable place, in which a few organisms will survive, such as bacteria or fungi. So, for biohistorical records the Anthropocene related to the human footprint and with the massive extinction process is just one evidence that human beings can discontinue life. But unlike being a single blow with atomic force, just like the extinction of dinosaurs by a meteorite, this discontinuity process is slow, although not as slow as natural processes of extinction. Moreover, this process of extinction related to human footprint is evidence, just as Foucault said, that life and history are intertwined to such an extent that they interfere with each other. The difference between this type of interference with others registered across the last two centuries is that it not only transforms life but destroys it and not only drives away death but makes it present and manifest.[16] And to be more precise, this interference is throwing life into extinction.

What theorists have identified as the new geological epoch called the Anthropocene, in biohistory terms, is only the record of one form in which life and history interfere with each other. According to this analysis, the effect of this interference, registered at the end of the eighteenth century related to the steam engine invention,[17] is visible through human activities that involve technologies which act contrary to life itself in some ways. And probably there is no other interference contrary to life more important than the loss of biodiversity or, in other words, the extinction of species. This certainty is not only acquired by fossil collecting evidence but by the information collected from certain populations in specific periods of time. With this information collected and recorded through the population census, it is possible to know if the number of individuals has increased or decreased. If the number of individuals continues decreasing, probably the population and then the species will be critically endangered. And at a certain point of decreasing, the species will be considered extinct in the wild, with a probability of surviving in captivity. If the conservation efforts are not enough, then the remaining specimens will die and with them the species will become extinct.[18] So, in the best conservation scenario, it is possible to design multiple strategies for saving endangered species with information collected through census. In consequence, the simple fact of collecting information about different species, that are in danger or not, contributes to the work of conservation and preservation of species.[19] Even more, following Raffi Youatt's hypothesis, "a biodiversity census will help construct new ideas of a multi-layered and multi-species global community."[20] In this case, it is not just about conservation, but about changing the community's perspective by relating biopower to ecology. In any sense, conservation will come when the community becomes a plurality of biosocial collectivities, allowing another kind of relations between humans and nonhumans that could "transform existing forms of political identity,"[21] in transforming the very sense of the human "grounding it relative to other species and to the wide variety of local ecosystems that make up the global ecosystem"[22]; produce the so called bioregional identities responsible for their local environments "since they involve a greater sensitivity to the places in which humans live and to the nonhumans on which we depend";[23] and promote local social movements once certain species is considered threatened or endangered according to the information collected by census. Following these three points, Youatt concludes:

> ...the global biodiversity census constitutes in human minds the category of species with which we necessarily have relations and, arguably, to which we have responsibilities given their sentience, conativity and capacity for interior experience. It highlights our embeddedness in the ecosystems in which we always participate and have effects. And, it creates a sense that the diversity of life-forms is both an ethical good and a prerequisite for long-term sustainability of human and planetary life.[24]

However, Youatt's optimistic view of census, whose major effect is to produce a multi-species global community, acquiring the biosocial collectivity form, or in biopower terms, to foster life through the management of populations, species and ecosystems, the biodiversity census and the lists that focus on identifying threatened species produce, as Irus Braverman explains, one fundamental biopolitical issue:

> ...the various conservations lists differ in their perspective on what is most important about life and thus on what is most worth saving-whether rarity in numbers, unique territorial configurations, or evolutionary (phylogenetic) variation. Rather than a bifurcated understanding of life versus death, then, conservation lists parse the life

> of species into complex orderings according to their extinction risks, effectively establishing a gradation of animal bodies according to the perceived worthiness of their lives. Even among those species who are deemed threatened, categories and criteria thus prioritize the ones who are perceived to be *the most threatened of all*: those whose lives are even more, and finally most, worth saving.[25]

Considering Braverman's experience working with lists,[26] it is possible to say that Youatt's argument does not consider the biopolitical issue that brings the biodiversity census in its core, that is to say, in one hand, that the principle of selection is applied on a living being, either an individual, a population or a species, that is worth saving among others that are unworthy no matter how much a biosocial collectivity is formed and, in other hand, that lists generated from the censuses become ineffective when faced with economic costs of management.[27] But through the biodiversity census and the lists of endangered species, just as the IUCN Red List, that are collecting information about other species decreasing in number for anthropogenic reasons, can make visible the exercise of a power that throws life into death and extinction. Finally, this is the evidence that the pressures generated by the processes of history, specifically human history, interfere with the movements of life.

3. From Biohistory to Biopoetics: Avoiding Death and Extinction

> Yet the lesson of Foucault's final lecture is that "another world" may only be reached through an "other life", through making one's life otherwise than it was.
>
> —Sergei Prozorov, From Biopolitics to Biopoetics and Back Again

Something remarkable in Youatt's argument, no matter if the proposal seems utopian or difficult to implement,[28] is the opening of a possibility that would seek to change the interference produced by the processes of human history over the movements of life. That is the reason why Youatt resorts to the use of the new conception of biosocial collectivity, that make no difference between humans and nonhumans entities. Moreover, biosociality, just as the definition given by Paul Rabinow, implies a deeper relation between nature and culture: "…in biosociality, nature will be modeled on culture understood as practice. Nature will be known and remade through technique and will finally become artificial, just as culture becomes natural. Were such a project to be brought to fruition, it would stand as the basis for overcoming nature/culture split."[29] This definition is an example of how to imagine another form of interference, in this case between nature and culture. Rabinow specifies that biosociality could be the way to overcome the nature/culture split (just as Youatt wants to overcome the human/nonhuman split), but in strict sense, he is only ordering another form of the interference in which technique plays a main role in the artificialization of nature and culturalization of nature processes. In this case, biosociality is a name to describe how the pressures of movements of nature and the processes of culture interfere with each other. This is what in Foucauldian terms, can be known as biohistory, which focuses on the life/history split. So, biosociality is just another aspect of biohistory that can be related to the transformation of nature/life through the agent known as power/knowledge. This biosocial perspective more than overcomes the nature/culture split, is working with and through the split itself. In this form of ordering and working with the split, technique and technology play a leading role in the new nature/culture relationship, because it implies a technological remake of nature and culture (culture is a technological invention and in a continuous reinvention). In consequence, biosociality, just like biopolitics, seeks to transform nature/life and culture/history through different technologies of power and biopower. But, what if, contrary to biosociality, instead of working with the nature/culture and life/history splits, another way could be working with the interference itself? If biohistory can realize the historical interaction in which the interference occurs, then how can the interference be changed? If biohistory relates the past events and, above all, the effects produced by the interference between nature/life and culture/history, then which field using those effects can transform it? Foucault used biopoetics to point out some kind of resistance to biopolitics, as follows:

> Biopoetics would be justified because it is indeed a sort of personal fabrication of one's own life (note that in these arts the question often arises of whether or not an act is beautiful). One could thus follow the problem of sexual conduct: biopoetics where it is a matter of the aesthetic-moral conduct of individual existence; biopolitics where it is a matter of the normalization of sexual conducts according to what is considered politically as the requirement of a population.[30]

It is important to highlight that Foucault used this term only once in a note during the lecture course at the Collège de France *Subjectivity and Truth*, while he was explaining the concept of *bios* and its relation to technologies that allows it to give form a certain life. All of this in the historical frame of Greek antiquity. That's why Foucault refers to biopoetics as a personal fabrication of one's own life and a matter of the aesthetic-moral conduct of individual existence. Through this personal fabrication one can resist politics aimed to normalize the population's body. However, in addition to being a resistance to biopolitics, biopoetics can set up a different pressure over biohistory. In this case, biopoetics can be the agent that could transform the interference between life and history (or Nature and Culture), recorded by biohistory. Whether biohistory looks back to the past, biopoetics looks to the future. In the Aristotelian sense, biohistory relates the interference that happened in the past, while biopoetics relates the kind of interference that could happen. Following Aristotle, biopoetics could relate what interferences are possible (probability and necessity). Moreover, biopoetics does not focus only on the pressures that the *processes* of history interfere with the *movements* of life, but rather changing the interference itself could modify the relation between life and history, acquiring a new form in which the *processes* of life become a *movement* of history. In consequence, biopoetics do not only operate as an aesthetic-moral resistance promoting individual conduct against biopolitics, but also as a way to relate the possible interferences between life and history that would happen if the processes of life become movements of history.

To understand the link between biopoetics with biohistory is necessary to do a review of biopoetics' history in a deeper sense, because before and after Foucault, the concept has been used in different theoretical fields: from sociobiology to biosemiotics and from literature to antiquity studies. Reviewing these theoretical fields, the concept of biopoetics will acquire another content that allows recollecting heterogeneous tools by which it will be possible to relate the potential interferences between life and history that could happen in the near future. Rather than making a list of theoretical concepts waiting to be memorized, it is possible to imagine a gallery in which, while walking through, the concept is experienced aesthetically.

Biopoetics in biology. In the introduction to *Biopoetics* (1999), highlighting that natural selection has shaped the features of living species, including human behavior, social organization and art, Brett Cooke states that to understand "the profound and longstanding impulse to create and enjoy art we are well advised to attend to our evolutionary heritage."[31] However, despite many scientists and philosophers having been interested in the genetic influences of social behaviours, Cooke, realizing a lack of this interest considering the question of art, claims that the tools of sociobiology could help in "understanding of the centrality and persistence of the arts in human experience."[32] In close relation with evolutionary psychology, that places arts as a general activity traced in every human culture, this approach points out "that art is one of the products of natural selection because it is, at least in toto and over time, adaptive."[33] Moreover, it must seek *artistic universals* and *features* that resemble *humanity*. At this point, distinguishing from evolutionary aesthetics and bioaesthetics which focus on beauty of living things, Cooke who wants to study "the creative arts, human-made beauty,"[34] coins the concept of biopoetics derived from, on one hand, the Aristotelian definition of the word "poetics" in reference to art and, on other hand, the definition of the Greek verb that means *making*. In consequence, for Cooke biopoetics "refers to our impulse to create beauty."[35]

In *Biopoetics. Towards an Existential Ecology* (2016), partaking from the biosemiotic turn in life sciences, Andreas Weber explains, in first stance, that life is a sense-creating phenomenon and, in second stance, that meaning generation is to feel. Then, "organisms live in existential felt, highly biased realities which they experience through their selves."[36] This new approach that promotes a poetic of nature, could propose a change in the mechanistic, causal and genocentric perspective because organisms in producing themselves "generate a poetic standpoint."[37] For Weber, every organism considered as a subject creates itself at the same time when it encounters, interacts and interchanges meanings that will be correlatively interpreted. In this perspective, meaning and interpretation are the patterns followed by organisms according to their needs. Out of the anthropocentric perspective, the poetic approach assumes that meaning is nothing else than a physical matter. Then feeling, expression, value and meaning "are the center of a physics of organisms."[38] Within this frame, Weber defines the biopoetic approach, in which poiesis is understood as the relation between inwardness (perception of the world), its expression and the transformation of the individual–world relationships, as "a process of deepening connections with other life."[39] Instead a mechanistic view of life, this approach must take into account another weak linkage as the importance of epigenetics (every organism can influence the genetic expression) and the recent organism centered perspective in which an organism, being more than a machine, is considered a metabiome (multiple symbiotic relations with other entities), an ecosystem or a *super organism* than creates itself: "The most simple organism must be understood as a material system displaying the intention to maintain itself intact, to grow, to unfold and to make a fuller scope of life for itself."[40] Finally, according to the biopoetic paradigm of the living, Weber makes no distinction between human and nonhuman life. Moreover, human beings "are deeply interwoven into the material, mental and emotional exchange processes that all of the more-than-human world participates in."[41]

Biopoetics in literature. In first place, talking about Foucault's displacement from literature because of its humanistic character and anthropocentric perspective on arts and his interest of studying the relation between life and power, specifically in terms of biopolitics, Julieta Yelin, in *Biopoéticas para las biopolíticas. Una introducción* (2015), explains in Nietzschean perspective that "life is political because there is nothing in it than will to power, so too it could say that life is artistic while there is nothing in it than will to creation."[42] In this perspective, art and life are not separated. In consequence, art is not a human production or something, just like literature, that humanism could use to preserve "some reservoir of civilization."[43] For Yelin, if biopolitics is an exercise of power over life, then biopoetics would understand the artistic-creative life's dimension, that considers animality as the starting point of the artistic productions: "The writing animal, the painting animal, the living animal – that is to say, it creates – in we: They are the real object of biopoetic thought."[44] So, this approach proposed by Yelin invites to think artistic practices to understand biopoetics as a particular mode of aesthetic production. In second place, discussing the biopoetics in the evolutionary sense and challenging biopolitics analysed by Foucault as power over life, Vittoria Borsò in *Bio-poetics and Dynamic Multiplicity of Bios: How Literature Challenges the Politics, Economics and Sciences of Life* (2020), considers three perspectives to propose another approach characterized as an *affirmative biopolitics.* With Roberto Esposito's idea of politics of life (understood as "a faculty of self-organization that is immanent to the living"[45]), Georges Canguilhem's distinction between concepts *about* life and concepts *in* life (the last one referring to "maintain the vitality of life" and that "represents life's potentiality"[46]), and Donna Haraway's concept of situated knowledge (that instead of privileging the universal objective knowledge, pays attention in partial and particular locations and experiences) proposes to outline a form of aesthetics related to literature that pushes language to the extreme boundaries of itself, opening it up to the intra-relational role of outside and inside within the immanence of life."[47] The form of aesthetics sought by Borsò has the following traits: (1) in relation to materiality of writing, it must show the power of life and the living and "its interaction with political and epistemological practices"[48]; (2) leaving aside objective reductionism, "it addresses artistic techniques open to the crossing of discourses about life through that which exceeds life forms [...] and fosters its own dynamics"[49]; and, taking into account that literature has an active role, it can challenge biopolitics informing "of the excessive power of life."[50]

Biopoetics in antiquity studies. Starting from the distinction between biopoetics and biopolitics used once by Foucault, Sergei Prozorov in *From Biopolitics to Biopoetics and Back Again* (2022), tries to show that understanding the biopoetics of antiquity (understood as aesthetic-moral aspect displayed by individual existence) will complete Foucault's biopolitical notion (which refers to normalization politics of population). Through this exploration Prozorov wants to contribute to contemporary "discussions of affirmative biopolitics,"[51] bringing to the field of analysis the interconnections between biopolitics, governmental rationalities and techniques of the self or arts of living. Searching the articulation of biopolitics and biopoetics in Foucault's courses *Subjectivity and Truth* and *The Courage of Truth*, Prozorov focusses in analyzing the Stoic and Cynic forms of life to show, in one hand, how the articulation "opens governmental rationalities to various forms of reception, readjustment, or resistance by the subject,"[52] especially in the Stoic discourses; and, in other hand, how it constituted "a form of life through the confrontation with the existing world,"[53] in cynical practices. The Stoic discourse related to sexual practices allows us to understand the relation between discourses of truth and the constitution of the subject. However, this relation could be comprehended through the analysis of the techniques of the self "by which one analyzes, evaluates and transforms one's existence."[54] In this case, according to Prozorov, rather than functioning as a radical break with the social order, these techniques permitted, in first place, "reconciling the emerging Hellenistic code of behavior that valorized marriage with the fundamental values of the Greek ethics of *aphrodisia*";[55] in second place, made *livable* "the modes of governance in the Hellenistic monarchies,"[56] that could be understand in the biopolitical frame; and in third place, within the articulation between biopoetics and biopolitics, these practices have the same object: "one's life that could be formed, reformed, transformed, or perhaps deformed in both large-scale regulatory governance of populations and micro-level individual or group exercise that adapted, adjusted, or resisted the rationalities of this governance."[57] Considering the Cynic form of life analysed by Foucault, Prozorov points out that biopoetics could be subjectify in some "kind of politization that would not be reducible to the government of populations."[58] This *politization* is realized with the Cynic *parrhesia* (understood as truth telling) as an "act of *disobedience* in the face of all social norms and conventions."[59] Exposing these acts of truth-telling could show that it is possible to live otherwise and to transform the world in accordance with the truth. In first place, to live according to certain truth, Cynics took and altered the four attributes of truth exposed by platonic philosophy, particularly: truth is "something unconcealed, undistorted, straight, and sovereign."[60] A true life is public, but it does not mean that social conventions are recognized and ensured, because there is nothing bad in nature; a pure life does not refuse the physical and bodily domains, instead it seeks to pursuit poverty leaving aside the superficial and inessential "to arrive at the absolutely indispensable";[61] a straight life only follows the law of nature that promotes animality rather than social conventions and codes; finally, a sovereign life that is scandalous, dirty and impoverished takes the figure of a true king when "submits his life to tireless tests in order to be able to take care of others, to lead them out of their untruth by his own manifestation of the true life."[62] Acknowledging animality in this attributes of truth, it is easy to understand that Cynics did not try to exclude or simply dominate one's animal nature. "In contrast, the Cynics transform this negative foundation into a positive telos of human existence, whereby animality is not a given to be mastered or conquered within oneself but a *model* to be attained in one's existence through courageous practices of truth-telling that break with established way of living."[63] In consequence, if animality is a duty and an exercise, just said Foucault, then is "the manner in which this life unfolds in the self";[64] if this form of life exposed in public and being scandalous disseminates the truth, then this transformed existence "can eventually change the world at least."[65]

To conclude, walking through the gallery of biopoetics, it is possible to aesthetically experience different dimensions in the relationship between power and life. The reflection of biopoetics on biopolitics may invite us to consider biological perspectives about life to produce a certain form of resistance that at the same time manifests another form of interference between life and history. Starting from the point of biological perspective, artistic expression is rooted in life, but it cannot be reduced to the genetic level, just as sociobiology tries to explain. On the contrary, in biosemiotics perspective, the creative power (poiesis) is an effect of the interactions and interconnections between organisms. In these encounters, organisms produce themselves in the same way that they produce their environment. Within this frame, in which human beings are another organism among others, biopoetics changes the perspective in which genetics is relevant in the development of the organism, but it is not entirely determinant. The interaction between different meanings and interpretations produces the poetic aspect of life. If biopolitics is a form of power exerted over life which implies the control and regulation of meanings and interpretations realized between organisms, then biopoetics, that is not reduced to explore the genetic heritage in human artistic productions, would try to change meanings and interpretations in accordance with the creative power of life. At the same time, this biological discourse could change the interference between life and history, because it introduces a new form of understanding life and a new form to interact with different organisms in a more co-productive sense. This biopoetic approach could modify the relations of power that are throwing life and, in this case, organisms to death and, in the worst case, to extinction. And more than talking about the stories of extinction and doing the biohistorical record of extinction, biopoetics would talk about the proliferation of life and its multiple forms of creative power. Considering biopoetical perspective in literature studies, it is important to highlight that literature, changing aesthetics and feeling, can challenge biopolitical power over life when resituating the creative-artistic dimension of life and animality;

that is to say, biopoetics would search for the power of life that can resist the biopolitical control. And in this case, literature would relate to another form in which life and history interfere with each other, that goes according to affirmative biopolitics. Literature in biopoetical sense would expose power *of* life more than power *over* life. Finally, with the analysis of forms of life in antiquity (Stoic and Cynic), biopoetics would become another form of producing subjectivity in the present, using different techniques of the self that make *livable* the current process of extinction caused by global warming and anthropogenic activities. In this case, the discourses of extinction and global warming would form, reform, transform or deform one's life and subjectivity. Then biopolitics in an affirmative sense and biopoetics related to certain discourses of truth, which means the current discourses of climate change, would function together in a more creative way to avoid extinction. But in a radical sense, if biopoetics resituates animality as the manner in which *life unfolds in the self* and the mode for breaking the established ways of living, then, according to the new perspective centered on organisms, animality would acquire the model of organisms to disobey the conventions, codes and social norms that exert power of throwing organisms and species into death and extinction. In consequence, a biopoetics centered in animality understood as an organism (metabiome, ecosystem or super organism that creates itself in interactions with other organisms) would change the world at large.

Notes

1 Jenny Ash and Russell Barnes Directors, (2001), *Extinct,* Canal 4.
2 Number updated by International Union for Conservative Nature's Red List of Threatened Species.
3 This de-extinction fantasy, that displays a net of biopower linked with biotech interests, cannot avoid the problem that producing three species of (hybrid) Dire wolf it does not imply that the species is recovered. Actually, even when a few specimens are alive the species is considered extinct for the record. Just as Cary Wolfe has settled related with de-extinction projects: "Here, the tendency has been to focus the question of extinction and de-extinction on the brute material presence or absence of the animal's body and genetic material, when in fact, from the more supple and complex point of view that animates this collection, a species may *already* be said to be extinct even though the last survivors of its kind live on" (Cary Wolfe, "Foreword" in *Extinction Studies: Stories of Time, Death, and Generations,* edited by Deborah Bird Rose, Thom van Dooren, and Matthew Chrulew (New York: Columbia University Press, 2017), xi). In consequence, the paradox is evident: the exercise of biopower produces life bringing back some specimens which are *already* extinct.
4 See, for example, the article *Bio-histoire et bio-politique,* in which Foucault, denying the existence of human races, points out the complex relations established between different and heterogeneous 'clouds' of populations. In fact, this kind of biopolitics in which Foucault was thinking only used to manage and administer the population's

life. See: Michel Foucault, "Bio-histoire et bio-politique", in *Dits et écrits*, vol. III, 1976–1979, edited by Daniel Defert y François Ewald, 95–97 (París: Gallimard, 1994).

5 Quoted in Irus Braverman, "The Regulatory Life of Threatened Species Lists" in *Animals, Biopolitics, Law. Lively Legalities*, edited by Irus Braverman (New York: Routledge, 2016), 33.

6 Michel Foucault, "The Crisis of Medicine or the Crisis of Antimedicine?", *Foucault Studies* 1, (2004), 11, https://doi.org/10.22439/fs.v0i1.562

7 Ibidem.

8 Ibidem.

9 Ibidem.

10 Michel Foucault, "The Birth of Social Medicine", in *Power, Essential Works of Foucault 1954–1984*, ed. James Faubion (New York: The New Press, 2000), 134.

11 Ibid., 135.

12 Michel Foucault, *History of Sexuality: an introduction* (New York: Pantheon Books, 1978), 132.

13 Jonhatan Baillie, Director of Conservation Programmes, Zoological Society of London, after describing the extinction problem related, in one hand, to the anthropogenic footprint and, in other hand, to the possibility of doing something to save some species threatened, asks the following: Do they have the right to exist or do we have the right to drive them to extinction?" (Jonathan Baillie and Ellen Butcher, *Priceless or Worthless? The World's Most Threatened Species*, (United Kingdom: Zoological Society of London, 2012), 16).

14 Michel Foucault, *Society Must Be Defended. Lectures at the Collège de France, 1975–1976* (New York: Picador, 2003), 253.

15 For Joseph Pugliese, military violence, as a modality of biopolitics, does not consider only human life as something possible to be fostered or disallowed to death, but more-than-human life too. This non-anthropocentric approach is important "to disclose the operations of biopolitics on trees, soil, water, rocks, and other-than-human entities" (Joseph Pugliese, *Biopolitics of the More-Than-Human. Forensic Ecologies of Violence* (Durham: Duke University Press, 2020), 7).

16 Keep in mind that, as Foucault asserts, across the seventeenth century the pressure exerted by the biological on the historical changed by the fact that the economic development, the implementation of new techniques on agriculture fields, and the knowledge concerned with life produce, in one hand, that dead ceased "to torment life so directly" and, in other hand, that a form of control over life "averted some of the imminent risks of death" (Michel Foucault, *History of Sexuality*, 142).

17 Paul Crutzen, "The 'Anthropocene'", in *Earth System Science in the Anthropocene*, eds. Eckart Ehters & Thomas Krafft (Berlin, Heidelberg: Springer, 2006), 16.

18 See the case of the Mexican Wolf. By conservation efforts and the strategies of breeding in captivity, the species has passed from extinct in the wild to endangered species, although this triumph has taken more than thirty years.

19 See for example the Encyclopedia of Life (www.eol.org), whose main project is to gather all the information about species or life-forms in one place. Although the gathering of information does not mean to conserve or preserve (living) species, there is a small step to get to planning strategies of this type. Take as an example the so-called *A "neoenvirnomental" manifesto*, in which Rob Jackson and Peter Haff point out: "Millions of species don't stand a chance against the current onslaught of

habitat loss and degradation, climate change, pollution, and other factors. Archives such as Norway's "Doomsday" seed vault are important first steps in preserving life. We need living archives of many species, whole-genome sequences of most species, and a network of reserves and zoos to harbor remnants of the thousands of species that will go extinct in the wild each year" (Rob Jackson & Peter Haff, "A "neoenvironmental" manifesto", *The Ecological Society of America*, vol. 9(2), (2011), 87, https://doi.org/10.1890/1540-9295-9.2.87

20 Raffi Youatt, "Counting species: Biopower and the Global Biodiversity Census", *Environmental Values* 17, (2008), 405, https://doi.org/10.3197/096327108X343149

21 Ibid., 407.

22 Ibid., 408.

23 Ibidem.

24 Ibid., 409.

25 Braverman, "The regulatory Life", 33.

26 From the author, see another two articles: Irus Braverman, "Anticipating endangerment: the biopolitics of threatened species lists", *BioSocieties* 12, no. 1 (2017): 132–157. And Irus Braverman, "En-listing life. Red is the color of threatened species lists" in *Critical Animal Geographies*, eds. Gillespie K. and Collard R.C., (New York: Routledge, 2015).

27 Ibidem.

28 The creation of a biosocial collective or group often represents a problem, because human beings do not always want to be part of a collective, especially when it produces a specific identity that brings conflict with other identities to the point of rejecting all of them. Ian Hacking highlighted this conflict in *Genetics, biosocial groups and the future of identity*, published two years before Youatt's article. See: Ian Hacking, "Genetics, biosocial groups & the future of identity," *Daedalus* 135, no. 4 (2006): 81–95.

29 Paul Rabinow, "Artificiality and Enlightenment: From Sociobiology to Biosociality", in *Essays on the Anthropology of Reason*, (New Jersey: Princeton University Press, 1996), 99.

30 Michel Foucault, *Subjetivity and Truth: Lectures at the Collège de France 1980–1981* (Basingstoke: Palgrave Macmillan, 2017), 34n.

31 Brett Cooke, "Biopoetics: The New Synthesis" in *Biopoetics. Evolutionary Explorations in the Arts*, edited by Brett Cooke and Frederick Turner, (Lexington, Kentucky: ICUS, 1999), 3.

32 Ibidem.

33 Ibid., 4.

34 Ibid., 6.

35 Ibidem.

36 Andreas Weber, *Biopoetics. Towards an Existential Ecology* (Dordrecht: Springer, 2016), 4.

37 Ibidem.

38 Ibid., 5.

39 Ibidem.

40 Ibid., 6.

41 Ibid., 7.

42 Julieta Yelin, "Biopoéticas para las biopolíticas. Una introducción", *Uni(+di)versidades*, 3, no. 3 (2015): 4.

43 Ibidem.
44 Ibidem.
45 Vittoria Borsò, "Bio-poetics and Dynamic Multiplicity of Bios: How Literature Challenges the Politics, Economics and Sciences of Life", in *Life after Literature. Perspectives on Biopoetics in Literature and Theory*, edited by Zoltán Kulcsár-Szabó, Tamás Lénárt, Attila Simon & Roland Végsó, (Switzerland: Springer, 2020), 19.
46 Ibid., 20.
47 Ibidem.
48 Ibidem.
49 Ibid., 20–21.
50 Ibid. 21.
51 Sergei Prozorov, "From Biopolitics to Biopoetics and Back Again. On a Counterintuitive Continuity in Foucault's Thought" in *Biopolitics and Ancient Thought*, edited by Jussi Backmann and Antonio Cimino (Oxford: Oxford University Press, 2022), 169.
52 Ibid., 170.
53 Ibidem.
54 Ibid., 171.
55 Ibidem.
56 Ibid., 172.
57 Ibidem.
58 Ibid., 175.
59 Ibidem.
60 Ibid., 176.
61 Ibidem.
62 Ibid., 177.
63 Ibidem.
64 Ibidem.
65 Ibid., 178.

Part III

SYMBIOSIS OR THE PARTAKING OF LIVING WITH/IN/SIDE OTHERS

Chapter 5

SYMBIOSIS AND THE SYM-IMAGINARIES OF BECOMING-WITH

This chapter is about symbiosis and partner organisms. To start, I would like to go back to Haraway's question regarding extinction or the loss of partner organisms: What happens when a significant partner in our lives is no longer with us? From this question, others emerge: How much and whose lives are affected by the loss of others? How profound are these relationships we, as living beings, cultivate with one another? How do we practice caring with others in our relationships? How can we, through care and interest, practice the arts of making-world-with?

In the midst of the "symbiotic turn[1]" and the proposal of the Symbiocene[2] by philosopher Glenn Albretch, symbiosis is permeating as a theory to better understand the way we relate to other living beings. Throughout this chapter, I will be trying to argument for an ontology of sym-imaginaries, linked to three related concepts: symbiosis (relationships between organisms), symbiogenesis (how these relationships configure the natural evolving world of beings) and sympoiesis (making with). Sym-imaginaries, in brief, are the attachments (thoughts and feelings that mobilize us), sense of belonging and intense interests that are created in relationship with others, that are fundamental to response-ability (learning to respond) and to living with others in better, kinder, negotiated and caring ways.

In order to do this, I will be thinking-with the short story *Indra's Web* by Vandana Singh. A story that inspires me to think with others about not only symbiosis, but what being a part of something (nature, ecosystem, nurturing relationships, etc.) means, and how to cultivate a sense of belonging and attachment that may be a good scope through which we visualize other ways of relating to others and our surroundings. To start, I would like to think about it by calling out Isabelle Stengers idea on naturalistic wisdom, the basis for her proposal of ecology of practices, a kind of thinking and worlding that does not set progress as the only path forward. She proposes:

> I chose an open reference to the wisdom of naturalists who have learned to think in the presence of ongoing facts of destruction – with nothing beyond to justify it – who are able both to feel that the disappearance of any species is an irreparable loss, which makes our world poorer, and to accept the loss of so many species.[3]

This kind of thinking, for Stengers, allows a form of responsibility to unfold. Responsibility for her is not "a matter of who is 'truly' responsible, it is a matter of concern, and, as such, open to technical advice."[4] As I understand, her concept of responsibility it is deeply engaged with taking time to think, of opening our imaginations, of paying attention the best you can to a particular situation, not guarding yourself with generalizations, nor allowing yourself to delve on guilt if things do not go the way they were planned. Together, these two ideas, naturalistic wisdom and responsibility, lead Stengers to propose a sense and ethics (in lower case) of belonging, as well as creative thinking for a different kind of scientific thought. In the story *Indra's Web*, the character Mahua will be our guide through this kind of thinking.

Before we get there, this chapter is organized in three sections. The first section regards an exploration of how symbiogenesis, symbiosis and sympoiesis frameworks point towards understanding our associations to other living beings through a different ontology of relationships between organisms. The second section, followed more closely with the story of *Indra's Web*, will delve into how these sym-imaginaries can be understood as a matter of remaking spaces we live in, thinking with others and taking inspiration from symbiotic arrangements to re-assemble the spaces we share with others. Finally, I will close with a section of how these sym-imaginaries are created by attachments and interests that go further than using symbiosis as technology to better human lives, rather, these imaginaries can help us create a sense of belonging that allows to cultivate response-ability even in the face of loss and grief.

1. On Being a Part of Something/Someones: Symbiogenesis, Symbiosis and Sympoietic Arrangements

This section's objective is to clarify the sym-imaginaries through the introduction of three concepts that sustain it. The idea is to sketch the ontological switches this framework allows as a critique of the views regarding evolution and its biological framework. These are the biological discussions of what counts as a unit of life, the mechanisms or dynamics of evolution processes and, what then, is the biological object of study par excellence. I will not delve too deep into these long-lasting discussions, but I will try to make clear how the hegemonic stance in these debates tends towards hierarchical theories of competition over cooperation and a clearly delineated object of study based on individuality, oneness and centrality.

The concept symbiosis was defined by Heinrich Anton de Bary (1831–1888) referring to a long-term physical association between organisms of different species. There are many types of symbiosis, but for clarity purposes here I will only mention three kinds: (1) mutualistic symbiosis points towards a relationship in which both organisms receive a benefit from the other, a classical example is the relationship between corals and algae; algae give coral energy from photosynthesis and corals give algae protection inside the hard structures of their bodies; (2) commensalism symbiosis, in which one individual gets a benefit and the other is not harmed or made uncomfortable by the other, remora and sharks are a good example; and (3) parasitism, in which one species gets benefits while the other is damaged, we can think of the plethora of intestinal parasites.

Symbiosis can be seen as a range in which many types of relationships are possible, but these three elemental ones help us capture important sites within the spectrum. It is also important to say that all species responding to each other in symbiosis may, over time, create a different kind of symbiosis, that is, parasitism can evolve towards a commensalism symbiosis and vice versa.

Symbiogenesis is a biological and evolutionary theory grounded on empirical evidence since, at least, 1923 when Boris Kozo-Polyansky wrote the text *Symbiogenesis: a new principle of evolution.* I also understand it as an interesting science story where there are paradigm shifts, tense dialogues and discussions. Symbiogenesis is a theory that countered the hegemonic status of the Darwinian Natural Selection Theory[5] in many important senses, a central one is that the mechanisms for evolution (specifically hereditary mechanisms) are not only mutation and recombination, but also a process of symbiogenesis or association. Kozo-Polyansky proposed that the process of joining and/or fusion of two or more assumed individuals because of processes of symbiosis (relationships), is a fundamental mechanism to evolution.

This theory, as Kozo-Polyansky suggested, shows how much speciation is created by these relationships between species. It means, that as much as we have thought of evolution in the metaphor of a tree, with a central trunk and expanding, differentiating and bifurcating branches, there are also processes of communion, of joining of branches to create a complete new evolutionary branch,[6] or what Lynn Margulis and Dorion Sagan call anastomosis, the fusing of the branches of the phylogenetic tree.[7] Kozo-Polyansky explains this by calling for examples of the creation of organelles inside a cell, stating, "[a]t the same time, [many of] the cell's organelles are not a product of differentiation but instead are the result of compositeness, joining and incorporation from outside of the unit, of life."[8]

A classic example is the origin of eukaryotic cells (nucleated cells with differentiated organelles and cell membrane, like the cells of plants, animals, fungi and protoctists), called the endosymbiotic theory. At the time Kozo-Polyansky wrote about symbiogenesis, eukaryotic cells were considered the only real cells. Prokaryotic cells (bacteria-type cells with no nucleus and cell wall) were at his time not considered cells in a strict sense. The story goes that some bacteria ingested another and did not digest it fully, but made it part of itself, creating what we now known as organelles (mitochondria, nucleus, endoplasmic reticulum, cytoplasm and others).

Kozo-Polyansky showed evidence of many symbiogenetic moments in which complex organisms are created by the introduction of one into the other, creating structures that later on will aid the organism's survival and boost evolution. Even as Kozo-Polyansky's studies had much empirical evidence, his hypothesis took many years to be accepted in biological studies and communities, at least in the American continent, and when this happened, it was mostly thanks to Lynn Margulis.

Today symbiogenetic theory is mostly accepted, although the mechanisms and particularities are still being studied. This theory created a paradigm switch in the biological understanding of evolution, because it enlightened towards another process or method of speciation and showed the overabundance of focus on the mutation process of evolution and the centrality of the genome. It also created another paradigm change focusing on symbiosis, or as Margulis would later propose, that the object of inquiry in evolution is not the cell, the gene, or the genome, but the *relationship between* assumed individual organisms.

More so, Kozo-Polyansky confronted the idea of individual organisms, by calling them organisms as consortia, or composite organisms. In his words, multicellular organisms are not colonies but a "consortium, a collection of initially heterogenous units."[9] With this, Kozo-Polyansky was not only challenging individuality, but criticizing the mechanical theories in biology: "The composite organism, with its tissues and organs, clearly is not a machine. It is a system of heterogeneous, self-sufficient, more elementary, bioblastic, single-celled and multicellular organisms. It is a federation of organisms."[10]

Kozo-Polyansky suggested that there would be at least three categories of hereditary variation: (1) combination where new entities are formed through association of heterogenous reproductive cells while interbreeding (sexual reproduction); (2) mutations, where new entities are formed due to enigmatic or unknown causes; and (3) fusion, where new composite entities are formed due to association of heterogeneous organisms (coming together in a system, fusing parts like cytoplasm, nuclei, genome or metabolism) besides those that are mediated by gametes.[11] The addition of fusion, a symbiogenetic process, to the mechanisms of evolution suggests that symbiosis is fundamental for life diversification, variability and long-lasting change. The relationships between symbionts may create a symbiogenetic process or they may not, symbionts do not always create an association so profound (fusion) as to generate a new species, but it may.

Following Kozo-Polyansky's and others' developments of symbiogenetic theory, Lynn Margulis and Dorion Sagan proposed that symbiogenesis challenges the popular evolutionary theories that claim that organisms evolve due to accumulation of mutations (random and unpredictable permanent genetic changes that accumulate with time), as well as, the Modern Synthesis Theory based on the unification of Darwin's theory of gradual change with the genetic developments of Gregor Mendel and his concept of stasis.

Margulis and Sagan's argument is not so much that the accumulation of random mutations is wrong, but more so "oversimplified and incomplete: It dismisses the symbiotic forest for the genetic trees."[12] They are, then, not disagreeing with the evolutionary theories of Darwin and Lamarck. They do not think natural selection and acquired characteristics are completely off the mark, but that the acquisition of entire genomes by symbiogenetic processes determines the characteristics in question. The endosymbiotic theory of eukaryotic cells shows this clearly, "the acquisition and integration of alien genomes for the first time not only led to the earlies eukaryotes, it led to the first species in the history of life."[13]

Following Margulis and Sagan, bacteria do not have proper species, since they are not formed by the integration of symbionts. In this sense, only eukaryotic organisms can be thought through the concept of species. This means, the concept of species they propose challenges other ways of comprehension of the term and of processes derived from it like speciation:

> the 'phylogenetic' or 'evolutionary' or 'cladistic' concept of species is entirely wrongheaded, and its adoption interferes with understanding how species arise. The long-term symbiosis that led to species origin by symbiogenesis requires integration of at least two differently named organisms. No visible organism or group of organisms descended from a 'single common ancestor.'[14]

Symbiosis challenges what Margulis and Sagan call the myth of the independent individual or the idea of a self-contained, independent and autonomous individual; because it shows how much organisms depend on each other and how much we are not isolated but always becoming-with. A clear example on this is the fact that "the estimate is that any person is about ten percent dry weight symbioses. Most human symbionts, by weight, are the many types of bacteria that live in the colon."[15]

Understanding symbiosis as fundamental in evolution, also challenges the story of evolution as progress, this is, the idea that evolution is a linear and vertical oriented process of enhancement or becoming more complex organisms, they state: "the very term 'progress', with its moral overtones denotes a complex quantity that is unmeasurable and unassessible."[16] Margulis and Sagan insist that these concepts (individual and progress) should be abandoned as they have been constant and biased frameworks to understand the process of evolution. As such, they propose a change of language in the way we usually approach them:

> The language of evolutionary change is neither mathematics nor computer-generated morphology. Certainly, it's not statistics. Rather, natural history, ecology, genetics and metabolism must be supplemented with accurate knowledge of microbes. Microbial physiology, ecology, and protistology are essential to understand the evolutionary process.[17]

They suggest, this knowledge provides ample language to comprehend that the main sources missing in Darwinian variation, such as attraction, merger, fusion, incorporation, cohabitation, recombination, sensitivity, co-optation, acquisition, accommodation, perseverance and other capabilities of microbes (and other organisms) "are not irrelevant to the evolutionary processes. Far from it. The incorporation of "foreign" genomes has been central to the evolutionary processes throughout the land and circuitous history of life."[18] In this sense, as they state, "evolution is a science of connection."[19]

Following Margulis and Sagans' proposal, Haraway offers the term sympoiesis, that means to make with. Sympoiesis points to the fact that all organisms in this planet make and are made by their relationships. Haraway develops a sense of sympoiesis based on the concept of Beth Dempster, who proposed the term first, but understood it as a process of co-creation occurring at a cultural and ecosystemic level. Haraway linked the concept towards collectively producing systems that do not have defined spatial or temporal boundaries and have exceptional potential for change. Sympoiesis, Haraway suggests, is in friction or generative unfolding with autopoiesis (proposed by Humberto Maturana and Francisco Varela in their book *De máquinas y seres vivos. Autopoiesis: la organización de lo vivo*) where the self-producing is autonomous, with temporal/spatial boundaries and with centrally controlled and homeostatic processes that make them predictable.

In Haraway's words: "Sympoiesis is a word proper to complex, dynamic, responsive, situated, historical systems. It's a word for worlding with, in company."[20] Sympoiesis challenges the notion of individual, it destabilizes the notions of cell, gen and organism. And it shows that any organism is always one and many, quasi-individual, quasi-social. By doing so, it switches the unit of life as an object of analysis and proposes the object as the multifactorial and dynamic relationships between beings. In this sense, sympoietic entities, as Haraway calls them, could be thought of as holobionts: "This is decidedly not the same as One or Individual. Rather polytemporal, polyspatial knottings, holobionts hold together contingently and dynamically, engaging other holobionts in complex patternings."[21] Sympoietic entities and holobionts are what Margulis and Sagan call symbiotic assemblages (and later on, symbiogenomes), and what Kozo-Polyansky called composit organisms. All of them point to a profoundly related web of organisms in sympoiesis (making-with).

In summary, the sym-imaginaries proposed here through the exploration of symbiogenesis, symbiosis and sympoiesis, is a work in progress of understanding and learning from an ontology of relatedness that involves:

1. A critique of the Individual and Progress ontologies in science that derive from Neo-Darwinism[22] and Genocentrism,[23] accompanied by the proposal of symbiotic relationship and arrangements, composite organisms or holobionts, as fundamental to evolution.
2. A proposition to understand evolutionary processes with a different microscopic and macroscopic lens and language. This means, challenging the idea that evolution occurs only by mutations and allowing other concepts to come into play, especially fusion as proposed by Kozo-Polyansky. This, in turn, creates an understanding that evolution is more than genetics and valuing the importance that environment and relationships have in this process. In this sense, evolution is not only about competition versus cooperation, but rather it is about association, integration, consortium, novelty and creation. All of which are possible thanks to the relationships (symbiosis) between organisms: long-term relationships that create new capabilities (symbiogenesis) or shorter lived, grounded relationships of making with (sympoiesis).
3. Learning from the world as a multifactorial and dynamic understanding of the web of life, moving from the idea of phylogenetic trees to a forest of genomes interacting with each other, that creates the world we all live in.

We must keep in mind though, especially in the midst of the symbiotic turn, that symbiosis as well as many other biological theories may be used in many ways in the times of heavy biopolitics and the advancement of biotechnologies. Haraway suggests, that those who work with the "symbiogenetic imaginations and materialities"[24] can now use and expand these imaginations by the use of several technologies that have been developed, such as "all the powerful cyborg tools of the late century molecular and ultrastructural biological revolutions, including electron microscopes, nucleic acid sequencers, immunoassay techniques, immense and comparative genomic and proteomic databases and more."[25]

These techniques create a subject of study not seen as a simple individual but as a subject composed by several interrelations and interdependencies, and also, create new frameworks through which they try to understand the depth of the symbiogenetic processes. In a sense, Kozo-Polyansky was spot on when he stated: "In this principle of internal, endosomatic and intracellular struggle, biology can possibly find a key to understanding the causes of mutations and other hereditary phenomena"[26]; because now, paired with the technologies Haraway points out, the symbiotic turn has taken momentum and is developing side by side with developmental evolutionary perspectives (developmental evolution or evo-devo is the study of how developmental processes during an organisms life is related to evolutionary changes between generations) as well as epigenetics (the study of changes in organisms by the modification of the expression of genes rather than changes in their genetic code itself, generally driven by changes in the environment).

In the Symbiocene era, it is important to remember that even if the symbiogenetic imaginaries and materialities seem to be on full speed development, this does not necessarily mean that the focus has been to discuss the topic of the immensely complex interconnectedness of all beings on Earth, guiding towards better ways of relating with others. These Symbiocene imaginaries are mostly used in biotechnology and bioengineering developments in their most productive and anthropocentric sense.

Even though Margulis and Sagan propose a symbiotic theory that heavily critiques anthropocentric views, scientific knowledge of symbiotic arrangements can be (and is generally) used anthropocentrically. We probably all know the benefits of probiotics in our bowels and bodies in general, and industry marketing has exacerbated the *need* for them through practices of industrial production of model organisms (genetically controlled, isolated from other diverse beings, environmentally enclosed, rapidly reproducing) used for a purpose, always and only benefiting directly or indirectly human beings.

With these precautions in mind, sym-imaginaries stand as a counterweight to anthropocentrically focused uses of symbiosis. It urges us to think deeply about questions like: Why is it important to understand how we are related to others in the vast web of life, to switch from the metaphor of phylogenetic trees towards a forest of genomes instead of a linear escalator towards progress of which the human form is the most complex? How can we use sym-imaginaries to counteract the use of symbiosis as an anthropocentric biotechnology in dense exercises of biopower? Can sym-imaginaries guide us as a basis of grounded practices of living and relating with others in the arts of cultivating what Haraway calls response-ability? How can we learn to be less deadly and less parasitic partners? How can we re-assemble environments in a sympoietic way to live and die well with our partner beings? What kind of relationships could be cultivated through response-ability to flourish with others on a damaged planet?

2. On the Response-abilities and Responsibilities of World-Making Symbionts, in *Indra's Web*

> Once again, we are in a knot of species coshaping one another in layers of reciprocating complexity all the way down. Response and respect are possible only in those knots, with actual animals and people looking back at each other; sticky with all their muddled stories.
>
> —Donna Haraway, *When Species Meet*

> At the lab door there was the familiar sign that Salman had installed by encouraging vivid green moss to grow on an earthboard so that it spelled out the motto of the Biomimetic Energy Materials Lab – To learn from Nature, not to Exploit Her. His colleagues had teased Salman about the irony of exploiting a moss to write these noble words, but the moss seemed to have decided things for itself. It spelled 'To learn from Natur, not to Explod,' which still made her smile.
>
> —Vandana Singh, Ambiguity Machines

Indra's Web, by Vandana Singh proposes a scenario of a sustainable community called Ashapur, the "City of Hope,"[27] designed to contain and, in a sense, avoid the damage caused by natural disasters from climate change. The principal character in the story, a scientist called Mahua, is investigating an issue arising from the Suntowers, the communities' energy storage and production centers. She is one of the head scientists in the town's project of sustainable energies, she is also a passionate person when it comes to nature's patterns and energy cycles.

Mahua, along with a team of other scientists and engineers, designed the Suntowers by creating an energy technology they call Suryanet, inspired by and using fungal-plant symbiotic networks, called myconet. These networks are, as she understands, a chemical language that help both the symbiont organisms survive by engaging in different bonding relationships: through the protection benefits of their communication (sharing information of environmental changes or potential dangers), photosynthetic (sharing energy produced by photosynthesis) and other metabolic symbiotic assemblages. Mahua and her team learned to decipher this language, as much as they can, and created the smart-energy grid they now lived with. But, at the beginning of the story, they start to fail:

> Above this vista rose the Suntowers like a surrealist's dream: four functioning, and the fifth under construction, their tops capped by sun-tracking petals of biomimetic material containing tiny, environmentally benign artificial cells, the suryons, that drank up the photons. The largest, oldest facility was Suntower I, now mysteriously moribund.[28]

Since many of the residents were originally climate refugees and, most of them come from drowned villages in Bangladesh, they understood the importance of making their town less invading for the nature around it and much more integrated with it, they "were traditionally energy efficient, living in clusters, throwing away nothing and reusing everything."[29] Ashapur was carefully designed to integrate itself into their environment as much as possible, this meant they tried to keep the original pattern of the slum, planted a green carpeted town with rooftop gardens with solar panels and corridors of native trees like neem, khejri and gulmohar.

This arrangement was meant to help maintain biodiversity as much as provide psychological benefits of being close to nature. They tried to maintain the slum pattern of the place, minimizing the risk of drowning and avoiding building pathways that would interfere with its original topology and movements of animals throughout the landscape, as well as protect the settlement from overheating and providing seasonal supplies of food for the townspeople and animals.

> The narrow roads were not built on a rectangular pattern but instead curved, moving obligingly around an ancient peepul tree or dwelling. Her designers had kept the street pattern of the original slum but had improved on it, allowing room for people to congregate in front of this chai-house or in that niche, so that old women could gossip and mind the little ones, and the wandering cows and pariah dogs had room to rest.[30]

The landscape planning project of Ashapur was designed with a profound interest in living with other organisms by keeping some of the natural characteristics of the space and adapting the human-built spaces to fit into it, instead of against it. In a way, this design project is in tune with sym-imaginaries, in which not only the environment, but also the diverse beings whom they live amongst are considered, they are important, and they are not kept away from town. Even as they use a symbiotic relationship as technology for the generation of energy in their town; the design planning of the town, and the individuals in them, also give back to the network of relationships they live with.

Ashapur, then, could be thought of as being a project of response-ability. Which, for Haraway, means a process of paying attention to others and the environment, enough to learn to respond differently to the challenges of the symbiotic relationship they were faced with. Response-ability is not a secluded capability of humans, it is what critters do, they interact with each other in open and curious ways, they create symbiotic arrangements by a process of learning with; it is the way critters become-with symbionts. Hence, response-abilities must be cultivated, created on site, in "polite (political/ethical/in right relation)"[31] encounters with others.

In Haraway's terms: "Responsibility is a relationship crafted in intra-action through which entities, subjects and objects, come into being."[32] In this last quote, Haraway's term is close to Stengers proposition of responsibility in a minor key, "as a matter of pragmatic ethos"[33] where the responsibility we have is to pay attention as best we can, "to be as discerning, as discriminating as you can about a particular situation. [...] to decide in this particular case and not to obey the power of some more general reason."[34] In this sense, responsibility and response-ability are not guaranteed recipes for success or goodwill moral propositions, in always easy or beneficial encounters. More so, they are ways of relating to others and caring to interact with them enough to create better ways to respond, better ways to live and die.

This is clear in the story, as even as carefully planned Ashapur was, one of their energy grids, the first Suntower they ever built was dying. Mahua meets with the team to listen to their worries and proposals to repair the Suntower, as she listens she understands that not every one of them is comfortable with the energy producing system as they feel it is not controlled enough. While she hears them discussing she reminisces about the first time she saw them, nearly a decade ago, and she remembers the speech she delivered to them that day. Which helps grasp the essence of the project:

> I come to you with empty hands, [...]. What I can give you is a chance to be part of a revolution. A revolution that might just save our Earth from the climate crisis. One that comes up with not just new technologies but new ways to live that are more whole and deep and satisfying than anything you've known. [...]. The Ashapur project is like college, except that you have the chance to learn by doing, and to do more good than you ever could with the life now laid out for you.[35]

Mahua is interested in technologies inspired by nature's patterns, as we have seen, but she is also interested in finding, through them, *new ways to live*, which could be a way into the realm of response-ability. Learning from these patterns means not only creating technology but creating *new ways to respond* through technology, even, or especially, when technology fails to respond adequately. The revolution she speaks about is finding ways to live that are satisfying in that they would be living their lives working towards making others' lives better. Even if this revolution is not guaranteed, she still believes they would be doing much good to the world they live in, hence, she is sketching the philosophy of Ashapur project, a city of hope for them, for their lives and the ones who surround them, but also, in a sense, of hope towards the future. In the same tone, Mahua instructs the team to find creative ways to respond to the issue the Suntowers are having:

> You know how to check everything according to the protocols. Do that. Then come up with some wild theories and shake up the possibilities, but don't panic. We can survive a couple of power failures; it isn't as though we haven't grown up with them. The hospital seems to have enough power right now; so do most key-need places. Make sure backup generators are working and just wait.[36]

Mahua knows failures are a part of the project, they show the limits of the project as it is, but they do not limit the possibilities to respond in other ways. She insists on finding in those failures new knowledge of how to respond better; but also, she invites the team to stir up the possibilities, to think outside of the box, and not to stay only with what they already know. In this sense, there is a logic of response-ability in the philosophy of the construction of Ashapur. Mahua believes that technology is meant to be focused on living better lives and that its benefits are not only for humans. They are also learning to live with others, wild animals and trees as much as dogs and cows, designing places that allow symbiotic relationships to flourish. Ashapur is a human settlement, with human designed landscape, but it is not only designed for humans. They have expanded the concern for a myriad of beings and their sympoiesis arrangements, which brings new possibilities to symbiogenetic induced change and communion. In this sense, even as they are using technology inspired by other symbionts, they do so in a way to hold space for symbiosis to flourish and develop. Outside the human form of polis, cities and other hyper technological and isolated settlements; there are many kinds of technologies and designs of living quarters we can take inspiration from to learn new ways of terraforming sympoietically (making with, in relation, paying attention).

An example of this could be termites, who live in highly specialized and organized termitaria and who have excelled in the agriculture of fungi. Margulis and Sagan explain that everyday termites tend the fungi gardens, they prepare and fertilize the soil as well as clean any debris.[37] This agriculture, or domestication of fungi by the termites, is not a unidirectional relationship, it is mostly mutualistic and accompanied by others, like the protists in the termite's belly. Termites use other beings and are used by others, in this sense they are user and used, they create a world with fungi, as much as hold space (inside and outside their bodies) for symbiotic arrangements to flourish. Even if termites use the fungi and protists use the termites, this does not mean there are instrumental hierarchical relationships between them, they are all in the process of becoming-with. In that sense, as Haraway states:

> Instrumental intra-action itself is not the enemy; indeed, I will argue below that work, use and instrumentality are intrinsic to bodily webbed mortal earthly being and becoming. Unidirectional relations of use, ruled by practices of calculation and self-sure hierarchy, are quite another matter.[38]

Ashapur local action and design, as I see it, can help us realize how technology can be used in a sym-imaginaries framework to create better living spaces and conditions for, not only humans but also many other beings. It shows it is possible in a short to medium period of time to develop such technologies that aid us all to have better lives. Of course, not every symbiotic relationship makes the lives of both symbionts better all the while, but Ashapur shows we can find ways of being much less parasitic and much more mutualistic with other beings around us by practicing response-ability and by creating different relationships of use and interest in one another.

What can we do to create our own cities of hope? Maybe, by situating ourselves in the space we inhabit, we are better able to find practices to fashion these cities of hope, even by creating small-scale practices and cultivating response-ability with other beings and the environment. There are many local practices we can start doing on a daily basis, like: composting organic waste, planting and caring for endogenous plants instead of big grass fields and exotic plant gardens, cultivating fruits and flowers according to the environment we live in and sharing them with other beings, holding space for local animals to live and move through respecting their freedom to exist and live in the spaces we share, rethinking our consumption of plastic, metal, paper and other materials as well as the decisions we make around food and products we use on a daily basis and, the classic, recycling, reducing and reusing what we think of as waste.

This is not an extensive list of techniques, and they may sound very basic, but they can create positive change towards living and dying better with others, or as Haraway suggests: "The doings of situated, actual human beings matter"[39], because "a common livable world must be composed, bit by bit."[40] Sf narratives can also be inspiring towards thinking about technologies being used to better the environment around us to point towards healthier futures. Lastly, other than techniques and technology, it is important to consider the ontological switch towards sym-imaginaries. Which means not only valuing symbiosis, symbiogenesis and sympoiesis but to create through them a mindset interested in cultivating response-ability, requiring us to look at each other with respect, care and interest.

Through sym-imaginaries we are able to recognize both the beneficial and the non-beneficial relationships we create with others, the instrumental and unidirectional relationships, as well as the mutualistic ones. But in order to make these sym-imaginaries work for the ongoingness of life it is important to practice responding to and learn from each other. Response-ability is the term Haraway proposes for this ongoing task of responding, in her words it is about: "both absence and presence, killing and nurturing, living and dying – and remembering who lives and who dies and how in the string figures of natural/cultural history."[41]

3. Beings of Compost Tangled Webs of Relations, in *Indra's Web*

> If learning to think is learning to resist a future that presents itself as obvious, plausible and normal, we can not do so either by evoking an abstract future, from which everything subject to our disapproval has been swept aside or by referring to a distant cause that we could and should imagine to be free of any compromise. To resist a likely future in the present is to gamble that the present still provides substance for resistance.
>
> —Isabelle Stengers, *Cosmopolitics I*

Mahua's thought patterns and attachments are of interest in this section of the chapter because it is thanks to them that Ashapur city became the city of hope for those people and beings who lived there or settled there. Here I am referring to Stengers' sense of attachments, "what causes people to feel and think."[42] Attachments and interests are fundamental guiding elements for scientific and not scientific practices, technology, techniques and practices. If, in the midst of the symbiotic turn, we find ourselves thinking about how technologies derived from this particular framework can be used to create better conditions only for humans, it may be because the attachments to the relational spectrum (how to respond better to diverse others, and how to create a sense of belonging) of sym-imaginaries is being neglected.

Mahua's attachments rely heavily on the patterns of nature, to her, "the forest was where she got her best ideas; it was an eternal source of inspiration."[43] That is where she got the idea for the symbiotic technology of the Suntowers and the design of Ashapur. She also feels a deep sense of belonging to the forest: "she was just another animal: breath and flow, a kite on the wind, a deer running."[44]

For Isabelle Stengers, there is an important distinction to be made between being a part of and belonging:

> We are all social beings, parts of a society, and an easy way to produce an objective lowering of what we feel and think is to emphasize that what we claim as ours is not ours at all, but identifies us instead as a part of our society. In strong contrast to this, you do not belong without knowing that you belong.[45]

The distinction between being a part of and belonging, then, involves an effort to think and feel that you are not merely a part of something bigger, a group of animals, a forest, an evolutionary web; but that you belong somewhere or with someone, meaning that the relationships you build and nourish matter deeply in your sense of self and purpose. The sense of belonging is paired with a deep interest in Mahua to understand the world she lives in. This interest derives from curiosity and awareness of being an integral part of the present of others. Throughout the story, we learn that even as a child Mahua was interested and even obsessed with nature's patterns; these attachments of hers make her a kin observer of the world. In her nurturing of this observation, she is not finding borders or distinctions, but communion:

> When she was a child Mahua used to follow ants as they moved purposefully across the dirt floor of the room. She wanted to know where they were going in such a hurry, and whether their abrupt paces and frantically waving antennae had some hidden significance. Later she came to realize that the ants followed invisible trails across the floor – that the world was full of secret communication channels, like electric wires between poles that rose above the tenements. It was as though some inner sense within her had opened, because following this realization she was suddenly aware of walking through a tangled spider's web of relationships.[46]

In Mahua, this awareness of belonging and deep interest in patterns was not necessarily always beneficial throughout her life, at the young age of thirteen she developed apophenia: "becoming sensitive to networks and relationships, she saw connections everywhere, even when they weren't actually present."[47] So, she decided to train herself to distinguish between the real and the imagined connections by studying the world, specifically she came to understand these patterns better as she came across the term: "Energy! And its conservation law. All real systems were governed by fundamental laws that acted as constraints. That is how you differentiated apparent relationships from real ones."[48]

The flow of energy being to being (for instance plant and fungi) and being to environment, is key for her biomimetic instrument-beings and energy plants called Suntowers. In a sense, just as Margulis and Sagan thought with the Gaia hypothesis, there is an understanding that: "Life is Earthbound and cellular, but also a geological and solar phenomenon."[49] All beings on Earth are connected through energy exchanges and fluxes. Solar energy, from a distant star, is fundamental to many processes that living beings, including humans, of course, need for their survival and wellbeing.

Suntowers in Ashapur sustain many lives by providing energy to the town. They are dependent on the Sun and the symbiotic connections developed inside the towers. When one of the Suntowers starts to fail, panic arises for the whole community's wellbeing; but to Mahua, there is one being in particular who is fully dependent on the energy from the Sun Towers: her grandmother. Who, after a stroke, was laying on a hospital bed, sustained by electric technology. Mahua decides to revise the fifth tower, the new, unfinished tower they were installing. It was already physically connected to the grid, but still not fully functional. As she examines the tower, she perceives:

> At first she sees just the skeleton, but there, between the supports is a new, delicate, lace-like integument. The suryons are distributing themselves, filling in the empty spaces. Somehow the Suryanet has not only decided that Suntower 5 should be functional, but has allocated resources to it, which is why Suntower I – the oldest and least efficient – has shut itself down.[50]

Once she leaves the fifth tower, she reflects on the fact that the energy grid they created won't be able to sustain much of the energy needed in town as the first tower perishes and the fifth one becomes functional. The redistribution of suryons was not something they expected to happen as they expanded the grid. Mahua is faced with the imminent death of her grandmother a truly significant partner in her life:

> At last she goes back into the night, thinking about the networks in which she exists, and how tomorrow a major node on which her very

> life depends will be taken offline forever. She thinks of the forest on the ridge. The forest lives on because it accepts death – with every twig that falls, with every ant that meets its annihilation, a thousand life forms come into being. Danger walks there and so its denizen learn adaptation; here, too, we must rebuild ourselves, define ourselves anew with each loss, each encounter. She remembers a story the grandmother once told her about Indra's Web, the ultimate cosmic network in which every node mirrors the whole.[51]

This last reflection shows how much Mahua feels attached and convinced of the energy fluxes in nature and how much they belong to them and depend on them. She understands in this way the loss of her grandmother as part of the processes of living in a network of beings who die for others to flourish. She is reminded of the cycle of life and death as a fundamental part of the energy fluxes, redistributions and transformations. More so, as she believes, there is a rebuilding and a redefinition one goes through with each loss and each encounter.

The symbiotic arrangement of practices shown in this story are not failure or loss free, they are evolving modes of existence in a community built by interested people and diverse beings, seeking to live better lives together. Mahua's way of thinking is interesting because of her curious, scientific, naturalistic and pattern seeking understanding of life, as much as in the emotional, historical, situated and cultural realm. In this complexity of thought, she is able to accept the loss of a partner organism (her grandmother and the Suntowers) but not in an uncaring or detached way. Becoming-with, in this sense, means accepting the myriad relationships we belong to and that these attachments shape us in the loss or the encounter.

In a way, Mahua's thinking values what Stengers calls symbiotic agreement:

> Ecology [...] doesn't understand consensus but, at most, symbiosis, in which every protagonist is interested in the success of the other for its own reasons. The 'symbiotic agreement' is an event, a production of new immanent modes of existence, and not the recognition of a more powerful interest before which divergent interests have to bow down.[52]

Finding new immanent modes of existence is fundamental to a sym-imaginaries framework. In summary, sym-imaginaries can help us think with a deep sense of belonging to the world and with others, that is backed up or created by historical and evolutionary events that precede us. The sense of belonging can be a powerful attachment and interest to cultivate response-ability. As we practice care and interest, as we value the flourishing symbiotic arrangements we belong to, we are better able to think anew, to rebuild ourselves with each encounter and each loss, to recreate and reorient our practices, technologies, worlds and communities. Through them we may start to cultivate the creativity needed to respond in different ways and to make world with, to create sympoietic worlding practices. Or, as Haraway would suggest:

> The decisions and transformations so urgent in our times for learning again, or for the first time, how to become less deadly, more response-able, more attuned, more capable of surprise, more able to practice the arts of living and dying well in multispecies symbiosis, sympoiesis and synanimagenesis on a damaged planet, must be made without guarantees or the expectation of harmony with those who are not oneself – and not safely other, either.[53]

Notes

1 The symbiotic turn refers to the expansion of works about symbiosis in recent years. This turn does not have a single proposing author, more so if symbiosis once was a concept worked with in biology, now it has become involved in different studies and fields like literature, art, technology studies, philosophy, ecology, anthropology, and others. See: Scott F. Gilbert, Jan Sapp and Alfred I. Tauber, "A Symbiotic View of Life: We Have Never Been Individuals", *The Quarterly Review of Biology* 87, no. 4 (2012) 325-341.

2 Glenn Albrecht, "Exiting the Anthropocene and Entering the Symbiocene", Minding Nature 2, no. 1 (2016): 12–16.

3 Isabelle Stengers, "Introductory Notes on an Ecology of Practices", *Cultural Studies Review* 11, no 1 (2005): 186.

4 Ibid., 188.

5 This theory suggest that Natural Selection is the fundamental mechanism for evolution of species, which in simple terms means that organisms who are better adapted to a habitat will survive and reproduce more than those who are not well adapted.

6 Boris Mikhaylovich Kozo-Polyansky, Symbiogenesis: a new principle of evolution, eds. Victor Fet and Lynn Margullis. (Cambridge, Massachusetts and London: Harvard University press, 2010), 117.

7 Lynn Margulis and Dorion Sagan, *Acquiring Genomes: a Theory of the Origin of Species* (New York: Basic Books, 2002), 15.

8 Kozo-Polyansky, *Symbiogenesis*, 109.

9 Ibid., 110.

10 Ibid., 110.

11 Ibid., 113.

12 Margulis and Sagan, *Acquiring Genomes*, 201.

13 Ibid., 143.

14 Ibid., 7.

15 Ibid., 18.
16 Ibid., 22.
17 Ibid., 204.
18 Ibid., 205.
19 Ibid., 126.
20 Donna Haraway, *Staying with the trouble: Making kin in the Chthulucene* (Durham: Duke University Press, 2016), 58.
21 Ibid., 60.
22 Neo-Darwinism is the combination of Darwinian classical theory of evolution and Mendelian Genetics.
23 Genocentrism is a perspective that privileges the genes as the main drivers of evolution, development and functioning of every living being.
24 Ibid,. 63.
25 Ibidem.
26 Kozo-Polyansky, *Symbiogenesis*, 115.
27 Vandana Singh, *Ambiguity Machines and Other Stories* (Easthampton: Small Beer Press, 2018), 154.
28 Ibid., 147.
29 Ibid., 148.
30 Ibid., 148.
31 Donna Haraway, *When Species Meet* (Minneapolis, London: University of Minnesota Press, 2008), 42.
32 Ibid., 71.
33 Stengers, "Introductory notes on an Ecology of Practices", 188.
34 Ibid., 188.
35 Singh, *Ambiguity Machines*, 150.
36 Ibid., 150–151.
37 Margulis and Sagan, *Acquiring Genomes*, 107.
38 Haraway, *When Species Meet*, 71.
39 Haraway, *Staying with the trouble*, 55.
40 Ibid., 40.
41 Ibid., 28.
42 Stengers, "Introductory notes on an Ecology of Practices", 191.
43 Singh, *Ambiguity Machines*, 145.
44 Ibid., 145.
45 Stengers, "Introductory notes on an Ecology of Practices", 190.
46 Singh, *Ambiguity Machines*, 151.
47 Ibid., 152.
48 Ibid., 152–153.
49 Margulis and Sagan, *Acquiring Genomes*, 133.
50 Singh, *Ambiguity Machines*, 154.
51 Ibid., 154.
52 Isabelle Stengers, *Cosmopolitics I* (Minneapolis, London: University of Minnesota Press, 2010), 35.
53 Haraway, *Staying with the trouble*, 98.

Chapter 6

SYMBIOSIS: FROM BIOPOLITICS TO SYMBIOPOLITICS AND BEYOND

In *Triggers: Introducing the Technosphere*,[1] a series belonging to Haus der Kulturen der Welt (HKW) channel, Birgit Schneider analyses the consequences related to technological and communication networks with their metabolic entanglements. Exposing a piece of cable used in submarine telegraphy that once connected Europe and North America during the second half of twenty-first century until 1903, Schneider shows the relation between organic and inorganic, human technology and earthly forces, microorganisms and plastic (polyethylene). The transatlantic cable is a piece of evidence, in first place, of biocenosis or a community of different beings, such as corals and sea shells, in which a technical object becomes into a biotope, that is to say, a space where environmental conditions allow the development of living beings; in second place, of the interplay and entanglement of organic and inorganic natures, technologies and ecosystems; and in third place, of a media ecology constituted of technological, social, organic and earthly forces. So, a piece of cable dipped in tree's milk can not only resist the negative effects of saltwater but also become a biotope or milieu in which a biocenosis is possible.

This example given by Schneider shows the improbability of technology producing a space in which different beings can not only coexist but become into others. In this case, a random and non-calculated effect of technology allows to think, at least, in three dimensions: (1) that a biocenosis created by technology could be understood as a technobiocenosis; (2) that the very idea of biocenosis must contemplate the coexistence not only of organisms but symbionts; and (3) that technology could be like another symbiont.

Perhaps, the relationship between technology and symbiosis could be strange, because the first belongs to the inorganic, abiotic and artificial realm, while the second belongs to the organic, biotic and natural realm. So, how is it possible that something like symbiotic technology or technological symbiosis exist? At first glance, these expressions could be impossible, because in biological discourses, symbiosis is defined as the relation established between two different organisms belonging to different species. Then, technology being a thing cannot establish a symbiotic relationship with biological species. Perhaps, technology could instrumentalize different organisms to produce a specific form of symbiosis. But it does not mean that technology properly takes part in the symbiotic process. Accordingly, to this idea, how is it possible to explain the relation between technology and symbiosis? Recently, in an interview for the YouTube channel *Thale's Well*, Katherine Hayles, best known for her book titled *How We Became Posthuman*, explained human cognition as the integration of biological and computational symbionts. For Hayles it is important, just as scientific research has probed, to consider the biological symbiosis, because human beings cannot survive without their symbionts (like the microorganisms inhabiting in the guts). At the same time, symbiosis could be used as a technical term, just as in computational media, to designate the technological symbionts integrated into the control and command centers by which human beings survive in actual societies or in human bodies like smartphones. This relation between biological and computational symbionts, Hayles says, is bound to a deeper dimension called technosymbiogenesis. If technogenesis is the strong relation between humans and technology since the emergence of *Homo sapiens*, that occurred as a feedback loop in which humans change technology and technology changes humans, then technosymbiosis must consider the relations between biological and technological symbionts by which both humans are produced and changed, and by which survive.

Contrary to Schneider's explanation, Hayle's approach uses in an explicit way the notion of symbiosis to explain that humans, in technosymbiogenesis perspective, are symbiotic entities that make symbiosis with both biological and technical beings. In this case, technosymbiosis allows us to think about both biological and technical realms within human existence. Even more, technosymbiosis is the way by which human beings could become into other in relation to their symbionts. Given the fact that symbiosis is a process of life that has become relevant in the last two decades, it is important to analyse the biohistorical effects produced in the Age of Biopower. In this sense, the main objective of this chapter is to analyse the form in which biopower transforms life when it considers the biological process of symbiosis. To meet the objective, this chapter is divided into the following sections: *Humans as symbionts* describes the way in which human beings are understood as symbionts in the Age of Biopower; *Human's symbionts* considers the biohistorical conditions under which humans have made symbiosis with other species; *Human symbionts* characterizes technology as a human symbiont to explain other forms of coevolution; *The regulation of symbiotic life* focuses on explaining the biopolitical effects of discourses on symbiosis in the case of symbiotic engineering applied in the animal industry; and *From biopolitics to symbiopolitics* attempts to show

how the scientific discourse of symbiosis can configure a form of resistance to the power over life that instrumentalizes and exploits life itself.

1. Humans as Symbionts

Since Foucault stated, in biohistorical terms, that in the age of biopower along nineteenth century, human beings started to learn "what it meant to be a living species in a living world, to have a body, conditions of existence, probabilities of life, an individual and collective welfare, forces that could be modified, and a space in which they could be distributed in an optimal manner."[2] In this sense, humans are living beings, specifically animals related to other primates. In the last two centuries, this hypothesis has become a biological fact, gradually. Evolutionary biology has contributed to consolidating not only the laws of evolutionary processes, but also to affirming that human beings, being a species among others, have evolved by multiple, different and heterogeneous processes. According to recent biological approaches, the claim that *Homo sapiens* is an evolutionary process open to time is not in dispute. What has been in dispute are reductionist explanations, such as mechanicism[3] and genocentrism during the second half of the twentieth century. Genetic mechanisms are important to understand in some way the evolutionary process, but they are not the only mechanisms in place. Theorists on epigenetics have shown the relevance of epigenetic processes in ontogenesis, hereditary transmission and environmental relations. The new organism-centred perspective has given different arguments to support the idea that organisms are important for the evolutionary process, avoiding genetic reductionism and environmental determinism.[4] An organism is not only the physical (structure and organization) expression of genes, nor the result of adaptation mechanisms to the environment. Organisms can change by their interactions between internal and external organisms and can modify their environment. In this case, organisms (internal–external) and environment coevolve. These perspectives put into question genetic reductionism and environmental determinism and show the importance and relevance of other biological aspects. It is also worth highlighting the symbiotic perspective. Although it came into the scene during the 1960s with the research of Lynn Margulis, in recent decades, it has had a great stir in biological explanations, to the point of being considered one of the fundamental processes of evolution.

This approach not only questions genocentrism but the very idea of biological individuality, because, just like Scott F. Gilbert, Jan Sapp and Alfred I. Tauber claim, "[t]he discovery of symbiosis throughout the animal kingdom is fundamentally transforming the classical conception of an insular individuality into one in which interactive relationships among species blurs the boundaries of the organism and obscure the notion of essential identity."[5] Within this symbiotic view of life, humans, being animals, are symbiotic organisms that establish relations with other organisms, both outside and within themselves. Furthermore, to explain this complex interaction between different organisms, the term holobiont has been introduced, understood "as the anatomical term that describes the integrated organism comprising both host elements and persistent populations of symbionts."[6] After reviewing different perspectives of individuality (anatomical, developmental, physiological, immunological, genetic and evolutionary), the authors conclude: "Our bodies must be understood as holobionts whose anatomical, physiological, immunological, and developmental functions evolved in shared relationships of different species. Thus, the holobiont, with its integrated community of species, becomes a unit of natural selection whose evolutionary mechanisms suggest complexity hitherto largely unexplored."[7]

Within the symbiotic frame, following Foucault, it could be argued that, in the age of biopower, human beings are starting to learn what it meant to be a symbiont in a symbiotic world[8] and to have a symbiotic body hosting other symbiotic organisms. Now, the condition of existence of this type of symbiotic body is characterized by a complex relation with other microorganisms (such as gut microbiota) without which the body could not survive. Probabilities of life would be guaranteed by the proliferation of mutualistic organisms and not so much by parasitic ones. Just as Gilbert, Sapp and Tauber assert in the case of medicine: "first and foremost stands the challenge of elucidating the complex relationships between health, disease, and changes in the human microbiome. The interactions of host genome, symbionts, and diet become critical."[9] Related to the case of medicine, symbiotic approach changes the individual and collective perspectives of welfare. Instead of thinking of the individual and the collective separately, welfare must consider a new aspect: the collective individual or individual collective welfare. Whether it is the welfare of the individual in relations to its symbionts or the welfare of a symbiotic collectivity opened to another symbiotic relationship. If the major force of symbiosis lies in the relationship between symbionts for changing an individual or a collective it is necessary to modify the relationship itself. And finally, if planet Earth is a symbiotic space, then symbiotic forces must be driven and distributed across the planet in an optimal manner.

Biohistorical effects of symbiotic approach, applied to different fields from medicine to agriculture, will be observed in the coming years. However, the re-emergence of this approach in the last decades has triggered some changes and transformations in the exercise of biopower. A new type of body is emerging, that is being produced by the exercise of power. It's no longer just about body-machine and body-species, just as Foucault did, but a body-symbiont or a body-holobiont. This new form of power, without leaving aside the other two types of body, will produce bodies through the transformation of symbiotic relationships. For example, let us take the next statement: "A functioning human organism is a symbiotic system containing a multitude of microbial cells -bacteria, archaea, and fungi- without which the whole would be seriously dysfunctional and ultimately nonviable."[10] According to this statement, a body-symbiont could be improved in its functions by promoting relationships between the multitude of microbial cells. At the same time, it is possible to damage the body-symbiont affecting or in some way killing that multitude. So, if the exercise of power *makes live*, then it will improve the symbiotic relationship; but if it *throws to death*, then it will ignore or, in extremis, damage the relationship itself. A new field of contested forces has emerged with the introduction of symbiotic view of life, as well as a new production of bodies.

Certainly, symbiotic approach is not only applied to human life and human bodies, but to living beings in general. Moreover, this approach can contribute to biohistorical analysis, searching for the historical symbiotic relationships between humans and other species, both internal and external. This could trace a series of relations established between humans and symbionts that are not necessarily microbial species, but other animals.

2. Human's Symbionts

Is it possible to understand domestication as a form of symbiosis? And moreover, could symbiotic approach explain domestication in another way? If symbiosis is a question concerning relations and associations, what kind of relationships and associations do humans establish with other species? Answering these questions could contribute to the problem raised by Peter Sloterdijk about the history of domestication:

> Cultural historians have made it clear that, with settlement, the relationship between humans and animals came under new auspices. With the taming of the human being by the house, the epic of domestic animals [*Haustieren*] begins as well. Their attachment to human houses is, however, not only a matter of acts of taming, but also one of training and breeding.
>
> The human being and the domestic animal – the history of this monstrous cohabitation has yet to be presented in a suitable manner, not to mention the fact that philosophers up to the present day are still in denial concerning what they themselves should look for within

> this history. In only a few places is the veil of the philosophers' silence concerning the house, the human being and the animal as a biopolitical complex lifted, and what was then to be heard were dizzying references to problems that for the time being are too difficult for human beings.[11]

With settlement another history of domestication started for human beings, because they established important relationships and associations not only with animals but also with plants [...] and with their corresponding microbes. Besides, passing from nomadism to settlement, human beings, in most cases, were the beneficiaries of cohabitation. Human beings learned how to breed and train animals and plants. But at the same time, human beings began to domesticate themselves, that is to say that they learned very well how to breed and train themselves. This double sense of domestication must be understood in the light of symbiosis. Human beings, without knowing, became into a symbiont of different types of symbionts in the technological structure named house. By domestication, human beings acquired other microbes that nomadic life did not provide them with. In this sense, becoming a host of different external symbionts made human beings symbionts of other organisms both internally and externally.

Through the emergence or creation of human houses, the revolution of agriculture reached a turning point that has been seen and studied in other societies, such as the Termites' cities. In *Acquiring Genomes: A Theory of the Origin of Species*, Lynn Margulis and Dorion Sagan explain that certain type of termites that eventually lost their symbionts, protists and bacteria that fed on wood, "invented an agriculture where they cultivate a wood-digesting partner -a fungus (or perhaps many similar fungi) with the logical name *Termitomyces*. 'Higher' social termites domesticate these fungi, which they sow and reap."[12] The agricultural process invented by termites is as follows: "Busy higher termites loosen and prepare the soil, fertilize it, prune unruly fungal growth and other debris. Finally, at the end of the season, they harvest their crop."[13] Something relevant in this agriculture process is that termites "suppress fungal sexuality while they encourage growth of edible young 'shoots',"[14] just as human agriculturalists do with plants and animals. This comparison is not casual nor ironic, because it tries to show, contrasting two types of domestication, agriculture and symbiosis, how human beings can lose and gain symbionts by other means, just as termites have done. Throughout the example of termites, Margulis and Sagan conclude:

> The bacteria's, protist's, fungi's and termite's tendency to fuse to produce emergent structures is the same as ours, only they, in hundreds of species, urbanized at least 80 million years ago, and we have been at

> our fancy fusions for only some 8,000 years. The mergers and fusions that we feel are uniquely human in fact have many precedents in the nonhuman world. The termite farmers are merely a single example.[15]

3. Human Symbionts

Understanding the history of domestication in the light of symbiosis is not an attempt to naturalize human activities; instead, it is a strategy for recognizing the ways in which human beings establish relations and associations with other species across history and by different means. But, which *means*? What can be understood by *means*? If settlement was possible with the technical invention of houses that allowed different symbiotic relationships, then what is the status of human technology in the light of symbiosis? The answer to this question will provide evidence of the symbiotic history of human beings. In this case, just as Sloterdijk claimed in the essay titled *The Domestication of Being*, human history can be understood throughout "the history of things, materials, and symbionts"[16] related to human facts. How can things, materials and symbionts be brought together as a singularity related to human facts? The singularity that meets that requirement is technology.[17] It is easy to understand that technology is linked with things (tools and machines) and with different materials by which they are produced (from silex or flint to plastic). But how can technology be thought of as a symbiont, specifically as a human symbiont? One dimension of technology involves the production of tools and machines; the second dimension, in Sloterdijk's terms, refers to the production of humans throughout anthropotechnology, that is, the set of techniques applied to human existence (psychic and social) to optimize it; and the third dimension is linked to the technological process by which human beings come into being and become into others: technogenesis. In other words, just as Sloterdijk claims, human beings cannot be thought without the technological dimension. So far, technology has produced human beings just as they have produced it. This inseparable association or relationship has impacted both the evolution of human beings and the evolution of technology.[18] Across history, human beings and technology have coevolved. Is this proof of symbiotic relationships? Now that a symbiotic view of life has become the new perspective, it is not so strange that the relation, association and coevolution established between humans and technology can be understood in the light of symbiosis.

Recently, in the article titled *The Concept of Symbiosis Applied to the Human Technological Culture*, Theo Wobbes explains

> that the entanglement of the human and technology is to be considered as symbiogenesis, as a protracted symbiosis between respectively, the biotic and the abiotic components of the two partners as a tacit internalization

> of dealing with artifacts. With this long-lasting symbiogenetic process causing both physical and cognitive transformations in human's ancestors, nature has taken a turn and thereby gave the biological evolution a new direction with the emergence of the modern human with an artificial or prosthetic life form.[19]

For Wobbes, human beings, as an outcome of symbiogenetic relationships with technology, are part of the biotic systems that are entangled with abiotic systems (tools and machines).

> "Through the process of internalizing, these abiotic things have been given the opportunity to become a whole with a biological system in the capacity of natural artificial or a prosthetic being, the human. Just as, in nature, symbiotic relationships of biotic systems are necessary to ensure the existence of the partners. We find this exemplified indeed in humans where symbiogenesis of a biotic and abiotic systems has led to interdependence of both systems."[20]

In this case, human beings have evolved with technology while this relationship has ensured the survival of the partners (mutualistic symbiosis).

In this context, the biological term of symbiosis is used to explain the emergence and continuity of human existence related to technology. But, as Wobbes claims, this symbiogenetic process has resulted "in ecological problems such as global warming, climate change, and loss of biodiversity that ultimately may be threatening for all life forms."[21] Now, concludes Wobbes, that human beings as symbionts realize that they are part of a symbiotic planet, then it is time for human beings to "be in accordance with the symbiotic biosphere they live in and are part of."[22]

Wobbes' conclusion fits very well with Foucault's claim that in the age of biopower human beings started to learn how to be a living being in a living world. According to the symbiotic view of life, human beings are learning how to be a symbiont, both biological and technological, in a symbiotic world. Human beings are learning how to deal with their own technosymbiotic body, other symbiotic organisms (internal–external) and their own technosymbionts.

4. The Regulation of Symbiotic Life

The introduction of the symbiotic view of life has produced different effects among many fields, such as medicine, nutrition, farming and agriculture. These fields of knowledge understand and explain their symbiotic object, that will be transformed by the exercise of power in the long run. Mechanisms of power establish forms of regulation within a new expression of security. In this case, these fields of knowledge could modify the symbiotic force existing in every association and distribute it along the earthly space or, in other words, the biosphere. If for Foucault nineteenth century biopolitics were seeking to regulate biological mechanisms of a population, such as birth, health, reproduction, endemic diseases, among others, then the twenty-first-century biopolitics would seek to regulate symbiotic processes modifying relationships and associations between organisms not only in human populations but in more-than-human populations interconnected between species. Passing from an individual collectivity of a body to a collective individuality of an ecosystem, symbiotic processes will be regulated to foster life. If human beings will be and live in accordance with the symbiotic biosphere, as Wobbes suggests, it will only be possible when the exercise of power over life functions and operates entirely with the discourse of symbiosis. This symbiotic process becomes a historical movement that will transform life. But how will life be transformed in the light of symbiosis? What is the role of human beings in the transformation process?

In *Darwin Comes to Town*, Schilthuizen uses the example of ants and beavers to explain how these two species became *ecosystem engineers*, understanding this notion just as Clive G. Jones, John H. Lawton and Moshe Shachak have defined it: "Ecosystem engineers are organisms that directly or indirectly modulate the availability of resources (other than themselves) to other species, by causing physical state changes in biotic or abiotic material. In so doing they modify, maintain and/or create habitats."[23] These engineers build their habitats and transform their environments. Although construction and transformation benefit themselves, they can also benefit other species. In the first case, a long series of Mymercophiles get benefits in the ants' nests; and in the second case, the environmental enhancement carried out by beavers "create new niches for whole swathes of other species."[24] With this biological approach in which engineers plays a main role in the construction of niches and in the relationship between species, Schilthuizen identifies human beings as the ultimate ecosystem engineer: "And, as with any good ecosystem engineer, in so doing it has created niches for cohabiting animals and plants. Not myrmecophiles, but anthropophiles."[25] So, these anthropophiles could benefit, both directly and indirectly, from the human engineered ecosystems. Ecosystem engineering, as many species show, is a form in which relationships and associations between species are established. Could this approach be seen in the light of symbiosis?

To answer this question it is necessary to consider, first, the global scale reached by human activities; second, the effects of anthropogenic activities over not only an ecosystem but climate and atmosphere; third, the complex interactions between living beings known as holobionts; fourth, the interconnection of biological and technological symbionts; and fifth, biosecurity and bioregulation aimed to govern environment for controlling things, materials, and symbionts. These five traits show another exercise of power that takes as its object a certain type of life which is already in a milieu or environment. So, in order to transform life, it is necessary to control, modify, and alter the environmental conditions.[26] If ecosystem engineering creates habitats modulating the availability of resources to other species, what could be understood as symbiotic engineering?

In *Low-carbon cows: From microbial metabolism to the symbiotic planet*, analysing the relationship between cows and ruminal microbes to reduce bovine methane emissions because "Cows and other ruminants, like goats or sheep, live in symbiosis with myriad microbes that help them to digest grass but, in the process, also produce methane. Therefore, methane inhibition aims to intervene symbiotic relations between the host animal and its microbes,"[27] Andreas Folkers and Sven Opitz try to show how symbiotic relationships between organisms (symbionts, holobionts) are techno scientifically intervened. This type of intervention is referred to as symbiotic engineering, which is defined by the authors as "techniques that manipulate symbiotic relationships to repair or optimize life processes and ecosystems – symbiotic engineering often requires elaborate biotechnological measures but also everyday practices such as eating probiotics to improve gut health."[28] More than individuals, those techniques take holobiont as object and "the relation between interlaced set of living beings,"[29] specifically microbes. For these authors, the introduction of those techniques gives shape to a symbiopolitics that seeks not only to make live or foster life but to "restores ecological relations,"[30] altering symbiotic relationships. As a form of biopolitics, symbiopolitics exert power over bodies, populations and environment. According to Folkers and Opitz, in order to reduce methane emissions, symbiopolitics takes as objects the cow's body, microbes' population and planetary environment. However, it is important to highlight that the notion of environment is not only oriented to the external one, but also to the internal environment of a cow's body where microbes inhabit. Then, it is possible to say that the cow's environment and the microbes' environments are interconnected. So, for changing the cow's external environment altered by global warming, it is necessary to change the internal environment of its stomach.[31] In times of global warming, it is desirable to reduce methane emissions. And to do that, a new form of biopolitics has emerged. When it is not an option to stop cattle breeding despite the risks and dangers carried in the production process, a symbiotic view of life has provided an opportunity for controlling and regulating the negative effects over the environment. Emerging symbiopolitics with its respective techniques, such as symbiotic engineering, not only manipulates but design symbiotic processes. However, it functions and operates not so different from biopolitics, because they seek to establish regulating controls or a security device for reducing pollution in times of climate change. As Foucault once said, these controls do not aim to eliminate the identified problem, but only to reduce it or, in the best case, to reach the optimal limit. Symbiopolitics probes that biopower is not only an exercise of power over human life, but a power over life itself, that in recent times functions and operates, as a way of control and regulation, in the light of symbiosis. Moreover, it is possible to claim that symbiopower is not only invading but transforming life itself via environmental and symbiotic modifications and alterations.

5. From Biopolitics to Symbiopolitics

> Where there is power, there is resistance, and yet, or rather consequently, this resistance is never in position of exteriority in relation to power.
>
> —Michel Foucault, History of Sexuality

Symbiotic engineering, understood as technique is the most recent evidence that, in the first instance, power produces knowledge; in second instance, power over life functions and operates with biological knowledge; and in third instance, biopower amplifies its effects using the historical innovations and novelties produced by biology and, nowadays, in a wider range of life sciences. As political technology that orders, produces and controls bodies, populations and environments, this type of engineering, being a *natural* ability (just like ants and beavers do), becomes an artifact for regulating symbiotic processes of other species. However, this is just another way to exert power over life itself. If this exercise of power resembles the new hope in times of climate catastrophe, it is because mechanisms of power are changing from make live to restore life; from foster life to save and regenerate life itself.

Symbiopolitics should be joined to other historical forms of biopolitics that have functioned in accordance with different biological discourses, for example, molecular politics, thanatopolitics, anthropolitics[32] or genepower and genepolitics[33] that have emerged once genetics became the hegemonic biological knowledge field during the twentieth century. But far from only claiming that biology serves for the exercise of biopower, it is possible to find in biological discourses ways to resist the exercise of power. It is enough to consider the review of Jacques Ruffié's book, titled *De la biologie à la culture*, in which Foucault points out, in relation to the problem of human races and racism,

> that a rigorous scientific knowledge can have an immediate political significance in an age when repetitive, global condemnation of racism, combined with tolerance in actual fact, permits the maintenance of

> segregationist practices and of insidious 'scientific' endeavors like those of [Arthur] Jensen or the shameful UN resolution on Zionism. Rather than a rhetoric whose indignations shelter so many complicities, a filtering of the problem of races in scientific terms is indispensable.[34]

Following Ruffié, Foucault makes three important points about race: (1) something like human races do not exist, instead it is a "statical notion – a 'population'";[35] (2) genetic polymorphism of a population is useful; and (3) "that a population cannot be defined according to its apparent morphological characteristics,"[36] but through the study of blood markers. Hemato-typology "now authorizes the dissolution of the idea of human race."[37] Through this biological evidence, Foucault advertises not only that there were not races in human species, but conceives humanity "as 'clouds' of populations that are interwoven together and combine a genetic inheritance that is all the more valuable the more its polymorphism is accentuated."[38] The importance of Foucault's review is the conclusions reached in which, according to molecular biology evidence, there is a sort of rethinking both biohistory and biopolitics. Biohistory "would no longer be the unitary and mythological history of the human species across time,"[39] while biopolitics "would not be one of divisions, self-preservation, and hierarchies but of communication and polymorphism."[40]

Following Foucault's approach, biological discourses driven by scientific rigour can be politicized in order to change social, political and cultural relational schemes. Moving from oppression and violence to emancipation and connivance, this scientific discourse displays a biopolitical strategy that privileges communication and polymorphism between the *clouds* of populations. In this case, rather than segregate and divide, this biopolitics would seek to disperse not only the idea of racism but also the processes of racialization that occur in Western societies. Since races do not exist, as hemato-typology has shown, all other physiological assertions about race should be dismissed, and the hierarchy created, abandoned. Then, this form of biopolitics would establish different regulation controls for managing the life of populations, strengthening ties and encouraging mixing.

In biohistorical terms, the emergence of certain biological discourse, just like that of hemato-typology, shows another interference between life and history. As Foucault refers, rather than tracing unitary and mythological history of human species across time, as the myth of racial purity, biohistory tries to show the series of discontinuities and interruptions in which different populations have not only met but mixed at certain historical moments. In addition, identifying racism as result of sociopolitical relationships rather than a biological fact, biohistory would narrate, in first instance, the contextual circumstances in which racism has functionated and operated as a mechanism of oppression, discrimination, violence, and, in the worst case, extermination and, in second instance, the contextual circumstances in which a form of resistance has emerged in certain populations fighting for emancipations to establish other forms of connivance and coexistence. Thus, this biological discourse about races and mechanisms of racism not only takes part of a larger body of scientific knowledge, but it serves to modify the relations of power that shape a society. As Foucault has shown, rigorous scientific discourse that somehow brings to light the power of life can modify and transform relations of power and, regarding life, of biopower. In this case, biopower not only uses biological discourse to manage the life of populations, but instead biological discourses also transform the mechanisms of biopower.

Within Foucault's perspective, this biological discourse, understood as resistance to a specific historical problem such as racism and racialization processes, is an important key to transform the mechanisms of biopower that only take and instrumentalize life itself. Through this example, Foucault has shown how to create and design a strategy of resistance to a certain exercise of power considering both the discoveries and advances in biology and the power of life. In this case, instead of searching for purity and hierarchies, the power of life is related to communication, polymorphism and mix between populations, because they strengthen the processes of life. Thus, Foucault's perspective shows the way in which relations of power are not focused on exerting power *over* life, but in considering the power *of* life to design a strategy of resistance to certain historical and sociopolitical problems.

Despite the critique over Foucault's perspective on biopower and biopolitics, Catherine Malabou follows a similar path in the article titled *One Life Only: Biological Resistance, Political Resistance.* In the introduction, Malabou asserts:

> That a resistance to what is known today as biopower -the control, regulation, exploitation, and instrumentalization of the living being- might emerge from possibilities written into the structure of the living being itself, not from the philosophical concepts that tower over it; that there might be a biological resistance to the biopolitical; that the bio- might be viewed as a complex and contradictory authority, opposed to itself and referring to both the ideological vehicle of modern sovereignty and to that which holds it in check: this, apparently, has never been thought.[41]

To design a proposal that moves from biological resistance to political resistance, Malabou points out that the entry of life into political techniques is only unilateral, non-dialectical and non-reciprocal. Therefore, all discourses on life originating from biology are simply taken over by politics to carry out the control and regulation of individuals and the population. This structure, which makes it impossible to conceive of resistance from the biological to the political, must be replaced by one that, in principle, reveals the possibilities inscribed in the living being and, consequently, the bidirectionality, dialectic and reciprocity with which one can move from the biological to the political. Modern biology (molecular and cellular biology) provides categories that make it possible to think about and open up other forms of resistance.

First, epigenetics establishes two systems of hereditary information, both at the individual level (ontogenetic development) and at the species level (phylogenetic development). This establishes new links between the individual and the species. The epigenetic history of cellular development (memory and inheritance) consists of three stages: (1) ontogenetic development: individual development based on differential gene expression; (2) inheritance: transmission of information from one generation to the next; (3) environmental conditions: gene dominance depends on internal and external factors, indicating that genes react to environmental conditions, thereby generating a memory of changes caused by the environment. In this epigenetic framework, living beings are always different in their individual development thanks to the expression of their genes. Therefore, the information transmitted to other generations implies difference and also allows the transmission of their genetic memory to the environment. In this sense, a living creature cannot be reduced to a program that repeats itself indefinitely, both in the sense of individual or species; on the contrary, it is always an entity distinct and different from others of its species, with the capacity to transmit different and differentiated information to other generations.

Second, cloning, understood as a mode of reproduction carried out through the procreation and transfer of nuclei (without a sexual bond), demonstrates the reversibility and irreversibility of difference, especially in the process of cell division, in the case of reproductive cloning. Thus, cloning is, as such, the realized possibility of the reversibility of difference, since cloning, which does not mean making faithful and perfect copies, opens up another process in cell division. Regarding therapeutic cloning, the concept of regeneration comes into play. While it indicates the natural repair of the body or some part of it, it is an action that mammals have lost and that biotechnologies have recovered. Regeneration is linked to the discovery of stem cells (non-specialized cells conceived as totipotent, pluripotent and multipotent, with the capacity to repair, reform and regenerate damaged organs and tissues). These cells offer a perspective of the future, in the improvement of medical technologies, and a perspective of the past, in which regeneration appears as an ancient property connected to primitive animals. This vision coincides with the prehuman, that is, primitive animals that naturally regenerated and the posthuman, that is, human beings who have acquired regenerative function through biotechnology. Therefore, we must not overlook the fact that cloning is a historical fact made possible only by biotechnology. Beyond its sole purpose of instrumentalizing or manipulating life, it can also generate other forms of access to the possibilities of resistance inscribed within living beings.

Finally, for Malabou, resistance can be proposed on the basis of a new materialism that successfully reconciles the symbolic with the biological. This means that symbolic life, understood as spiritual life, life as a "work of art," or life as self-care, does not erase or simply obscure biological life or the natural dimension of the living being. It is appropriate to point out that symbolic life and biological life are directly correlated with *bios* (usually understood as human life) and *zoé* (understood as life that every living being shares). Symbolic life is to *bios* what biological life is to *zoé*. Thus, if the fundamental task of future materialism lies in reconciling symbolic life with biological life, then it implies reconciling *bios* with *zoé*. Malabou's contribution could well be defined as the transgression of the established boundaries between symbolic life and biological life, as well as the corresponding overcoming of dualism by making these modes of existence coincide.

If Foucault and Malabou could agree on anything, it would be their approach to biological discourses with which they think and rethink a different strategy to resist biopower and biopolitics that only instrumentalize life. Whether showing the power *of* life or the possibilities inscribed in the potential of the living being, hemato-typology, epigenetics, cloning and stem cells, being not only information belonging to scientific knowledge, offer an opportunity to change and transform power relationships. If these two approaches aim to design a strategy of political resistance and, at best, to transform the mechanisms of power, what could be thought of as a type of power *of* life inscribed in the biological process of symbiosis? What can be understood as symbiopower and symbiopolitics?

Recently, in the article titled *Immunity and Symbiopolitics*, Audronė Žukauskaitė has rethought biopolitics in light of theoretical advances of contemporary biology surrounding the immune system. Just like Foucault and Malabou, Žukauskaitė considers biological discourses to design another biopolitical strategy. During the twentieth century, immune system was considered a defence system whose main function was to identify the self and the non-self. And as a metaphor of biopolitics, the political realm used to identify and discriminate what was based on difference and otherness. Biopolitics used to function and operate as an immune system establishing boundaries and fighting that was identified as different or something like non-self. In terms of strategy, this way of acting was considered negative and reactive. However, the research on immunology has shown that those ways of acting are not the only ones. First, it must be admitted the immune system "develops throughout the organism's life";[42] second, it must be recognized that the immune system has more than one, negative and reactive, response: "in its exchange with the environment the organism normally reacts not in a defensive manner but in an assimilative one (eating, breathing). The immunological 'self' can also be nonreactive or silent in relation to the cells of other organisms, such as in the case of pregnancy."[43] Passing from different theories and hypothesis on immunology, Žukauskaitė explains immune system as an ecological network that acts not only in a reactive manner:

> As Tauber argues, ecological immunology does not mean that every individual is in relationship with the environment which is self-obvious. It means that every individual is in a relationship with its internal ecosystem and an external ecosystem. Every individual is a holobiont, composed of internal microbiome, and the resident symbionts living in the gut, skin, and on other surfaces. This internal ecology is responsible not only for normal physiology but also for the functioning of the immune system.[44]

This explanation goes further, because it changes the traditional opposition between self and non-self. Now, it is more precise to speak about multiple interactions between the internal and external ecosystems. And as Žukauskaitė refers: "It is important to point out that these interrelations evolve and change during the entire life span of the organism, modifying both internal holobiont and the external environment."[45] For Žukauskaitė, this new approach named symbiopolitics can be extended to a planetary scale "which connects all human and nonhuman agents,"[46] and, at the same time, can be claimed that "every biopolitics is always already a symbiopolitics, affecting the conditions of liveability of both human and non-human symbionts."[47] Finally, symbiopolitics, in first instance, can be characterized as a non-human form because humans do not live alone and isolated but accompanied by other living beings, including pathogens; in second instance, the self/non-self split has no sense and must be overcome, because "the 'other' is what makes 'my' immunity possible";[48] and in third instance, difference and heterogeneity are put in an affirmative way, then a new biopolitical model emerges that prioritizes "philosophical models that can conceptualize differences and heterogeneity in an affirmative way."[49]

To conclude, according to these three perspectives given by Foucault, Malabou and Žukauskaitė, it is important to highlight the importance and relevance of scientific knowledge, specifically the one focused on biological discourses, to modify biopolitical mechanisms and the biopolitical model itself. For doing this, it is necessary to pay attention to the recent research and theoretical advances in contemporary biology for bringing to light the power *of* life, the possibilities inscribed in living beings and the affirmative effects with which is possible to design a strategy to resist the harmful effects produced by an exercise of power that instrumentalizes and exploits life itself or that uses life and living beings to control life itself.

Reaching the time of biopolitics seen in the light of symbiosis, it is possible to assert that a new way of controlling and regulating life and biological processes has emerged. Symbiotic engineering shows the techniques and procedures by which a body, a population and the environment are affected using new mechanisms of power. However, it is important to consider that it is also possible to design a resistance following the biological discourse of symbiosis, that encourages and promotes different relationships between different organisms, specifically symbionts and holobionts. In this case, rather than functioning and operating through self–non-self split, symbiopolitics prioritize difference and heterogeneity. After all, difference strengthens life and biological processes. Whether it is the difference of interbreeding populations (Foucault), the difference achieved through epigenetic inheritance (Malabou), or the difference that makes the immune system possible (Žukauskaitė).

Finally, symbiopower and symbiopolitics must consider the role of technology as a symbiont, because rather than functioning in a mutualistic mode, it is *behaving* as a parasite. Just as Jan Zalasiewics, referring to the technosphere, has pointed out:

> Currently, the technosphere might be regarded as parasitic on the biosphere, altering conditions of planetary habitability. Obvious consequences include greatly increased (and accelerating) rates of extinction of species of plants and animals, and changes to climate and ocean chemistry that are largely deleterious to existing biological communities. These changes can in turn damage both the functioning of the biosphere and human populations. Ideally, therefore, humans should try to help the technosphere develop into a form that is more sustainable in the long term. Nevertheless, humans collectively have no choice but to keep the technosphere operative – because it is now indispensable to our collective existence.[50]

In this case, technosphere and technology, being indispensable to our collective and symbiotic existence, must be regulated. Considering that technology meets the characteristics of difference and heterogeneity, symbiopolitics must design a new world in which technology becomes a mutualistic symbiont by which other organisms could become into others.

Notes

1 Technosphere, according to Peter Haff refers to the novel sphere on the planet made of the human technology, adding to the classical spheres such as hydrosphere, biosphere, atmosphere and lithosphere.
2 Michel Foucault, *The History of Sexuality: an introduction*, (New York: Pantheon Books, 1978), 142.
3 Mechanicism is the philosophical and scientific doctrine that explains nature and living beings as machines.
4 Genetic reductionism creates a perspective in which from many available variables, the genetic one is the most important. And environmental determinism focuses on explaining the capacities and mechanisms of adaptation of organisms only by the environmental conditions.
5 Scott F. Gilbert, Jan Sapp and Alfred I. Tauber, "A Symbiotic View of Life: We Have Never Been Individuals", *Quarterly Review of Biology* 87, no. 4 (2012), 326.
6 Ibid., 327–328.
7 Ibid., 334.
8 For example, see: Lynn Margulis, *Symbiotic Planet: A New Look at Evolution* (New York: Basic Books), 1988.
9 Gilbert *et al.,* "A Symbiotic View of Life", 334.
10 John Dupré, "The Polygenomic Organism" in *Postgenomics. Perspectives on Biology after the Genome*, Ed. Sarah S. Richardson and Hallam Stevens (Durham and London: Duke University Press, 2015), 68.
11 Peter Sloterdijk, "Rules for the Human Park: A Response to Heidegger's "Letter on 'Humanism'", in *Not saved. Essays after Heidegger,* (Cambridge: Polity Press, 2017), 206.
12 Lynn Margulis and Dorion Sagan, *Acquiring Genomes: A Theory of the Origin of Species* (New York: Basic Books, 2002), 104–106.
13 Ibid., 106.
14 Ibid., 108.
15 Ibid., 111.
16 Peter Sloterdijk, "The Domestication of Being. The Clarification of the Clearing" in *Not saved. Essays after Heidegger,* (Cambridge: Polity Press, 2017), 99.
17 In *Wounded by Machines,* Sloterdijk identifies technological singularity as follows: "It turns out that there is a third dimension in the space of beings that escapes this ontological alternative: the machine, the artificial construction, that which is man-made. Between the realm of nature and the realm of grace there gapes an initially almost imperceptible fissure—the narrow domain of tools, works of art, and machines. As soon as one seriously makes the attempt, it becomes immediately evident that they were neither produced by God in the six days of creation nor were they allowed by the same God to later fall gracefully from heaven in wondrous breaches of orderly nature" (Peter Sloterdijk, "Wounded by Machines. Toward the Epochal Significance of the Most Recent Medical Technology" in *Not saved. Essays after Heidegger* (Cambridge: Polity Press, 2017), 229–230.
18 In the article *The impact of 'Anthropotechnology' on Human Evolution*, Sylvia Blad concludes the following: "The impact of anthropotechnology on our evolution cannot be assessed by genomic comparisons only. Although some evidence for an impact of 'anthropological' evolutionary history was found in our genes, the search highlighted the limitations of genetic comparative approach. Evolutionary significant differences

cannot only be found in changed genes, but also in changed genome architectures and dynamics, epigenetic patterns, and epi-phylogenetic memory" (Sylvia Blad, "The impact of 'Anthropotechnology' on Human Evolution", *Techné*, vol. 14(2), (2010), 25.

19 Theo Wobbes, "The Concept of Symbiosis Applied to the Human Technological Culture", *Technophany* 3, no. 1 (2025), 2.

20 Ibid., 19.

21 Ibid., 20.

22 Ibidem.

23 Clive G. Jones, John H. Lawton, and Moshe Shachak, "Organisms as Ecosystem Engineers", *Oikos* 69, (1994), 374.

24 Menno Schilthuizen, *Darwin Comes to Town: How the Urban Jungle Drives Evolution* (New York: Picador, 2018), 20.

25 Ibid., 22–23. If myrmecophiles is a concept that refers to the love for ants as well as the associations of ants with different species, then anthropophiles is a concept that seeks to refer to human associations with other species.

26 Following Foucault, Thomas Lemke explains that the notion of "environmentality captures a new dispositive of 'governing things', putting forward a distinctive ecological understanding of the environmental as both technological and natural" (Thomas Lemke, *The Government of Things. Foucault and the New Materialisms* (New York: New York University Press, 2021), 169).

27 Andreas Folkers and Sven Opitz, "Low-carbon cows: From microbial metabolism to the symbiotic planet", *Social Studies of Science* 52, no. 3 (2022), 331.

28 Ibidem.

29 Ibidem.

30 Ibidem.

31 In a more recent article, titled *Symbiotic engineering: insects, microbes, and the space of vector control*, considering the case of mosquitos that transmit viruses to humans, Opitz and Folkers insist on the necessity of controlling the external environment in order to control the internal environment of the mosquito. This symbiopolitics infects the mosquito's body with *Wolbachia* bacteria for both inhibiting reproduction and controlling populations. In this case, *Wolbachia* alters the mosquito's internal environment and at the same time its reproductive capacity. That symbiotic process results in a form of biocontrol and bioregulation of the virus. See: Sven Opitz and Andreas Folkers, "Symbiotic engineering: insects, microbes, and the space of vector control", BioSocieties 20 (2025) 481–503.

32 See: Thomas Lemke, "Beyond Foucault: From Biopolitics to the Government of Life," in Governmentality: Current Issues and Future Challenges, ed. Ulrich Bröckling, Susanne Krasmann, and Thomas Lemke (New York: Routledge, 2011).

33 See: Santiago Díaz, "El biopoder de la biotecnología o el biotecnopoder. Aportes para una bio(s)ética," *Ludus Vitalis* 19, no. 36 (2011).

34 Michel Foucault, "Bio-history and bio-politics", *Foucault Studies*, no.18 (2014), 128. Richard A. Lynch, translator of Foucault's article, mentions that Arthur R. Jensen "argued in the late 1960s and early 1970s that IQ was largely genetically determined, and that Blacks were therefore intellectually inferior to Whites. Ruffié discusses Jensen in a chapter entitled 'Clack racism and slavery' [...]; Ruffié cites several studies that refute Jensen's conclusions" (*Ibidem.*). About the UN resolution on Zionism, Lynch explains than Zionism was declared a form of racism and racial discrimination in 1975, but it was revoked in 1991.

35 Ibid., 129.
36 Ibidem.
37 Ibidem.
38 Ibidem.
39 Ibidem.
40 Ibidem.
41 Catherine Malabou, "One Life Only: Biological Resistance, Political Resistance", *Critical Inquiry* 42, (2016), 429.
42 Audronė Žukauskaitė, "Immunity and Symbiopolitics", *Athena* 17, (2022), 47.
43 Ibidem.
44 Ibid., 52.
45 Ibidem.
46 Ibid., 54.
47 Ibidem.
48 Ibid., 55.
49 Ibidem.
50 Jan Zalasewics, "The unbearable burden of the technosphere", *The UNESCO Courier* 2, no. 873 (2018), 17.

Part IV

BECOMING-INTO-OTHERS OR THE ATTENTIVE INTEREST IN CO-SHAPING OF KIN AND WORLD

Chapter 7

BECOMING-INTO-OTHERS, BECOMING EARTHSEED

At the beginning of the first chapter of this book, I used an epigraph from the novel *Parable of the Sower*, by Octavia Butler. Specifically, a quote her character Lauren Oya Olamina states on the Books of the Living, the fundamental volume of her philosophy Earthseed:

> All that you touch
> You change.
> All that you change
> Changes you.
> The only lasting truth
> Is Change.[1]

At that point, I did not explain the quote by following the novel. As an epigraph, it was meant to introduce the basic grounding for the seed of BIO to germinate. This last chapter, I will start by going back to the beginning, to the novels from which this quote comes from. First though, as this is the final sf chapter, I would like to make a re-vision of what we have learned so far, in order to create a more stable basis for understanding the proposal of BIO that is contained in this book, and which will be the centre of this chapter.

So far, we have explored different sf narratives following the questions that BIO inspired revolving three themes: transformation of life, extinction and symbiosis. Each chapter contains sf stories that allow us to think creatively about the themes, and about how the narratives may be related to the reality we live with. The exercise of imagination and curiosity sf authors embark on while writing is a beautiful craft that expands creativity in a myriad of ways; this is the gift sf writers give to us. They anchor us to reality in strange and unexpected ways, they allow us to think about reality with other lenses and tones.

I believe that by cultivating, practicing or exercising imaginative experiments of thought like those of sf, we may be better equipped to deal with the problems we face. This is because we allow ourselves to think outside the common places, names, geographies, laws, company, normativity, social frameworks, time frames and other conditions we are accustomed to.

By asking the question "What if?", as Ursula K. Le Guin suggested, sf authors are able to think and offer possibilities. Throughout this book, we have explored some of these questions: What if humans lived with another humanoid species as their symbiont? What if someone decides to breed powerful humans? What if a scientist modifies human genetic templates to create a better-equipped being for the world after the apocalypse he created? What if chickens survive and thrive in the world after humans have perished? What if humans are being saved from extinction by aliens who want to create a hybrid species? What if a community starts using symbiotic technology as inspiration for an energy grid?

All these questions are tied to our world, no matter how far away from our solar system they are set on, how far in the future, or how impossible they seem to us right now. If we pay close attention, we might learn from these narratives, not only what technologies and humans have been, fear or hope to be; but also, how do we want to live in the present and future, what world we would like to leave to future generations of humans and other living beings present and yet to come.

This last chapter will be trying to embody these thoughts through two experiences and attachments that are very personal to my own (but always shared) BIO. In order to do so, I will explore during the first section the philosophy of Earthseed contained in *Parable of the Sower* and *Parable of Talents*, by Octavia E. Butler. This philosophy has helped me in many ways, but in this chapter, I am exploring specifically how they inform and inspire the BIO proposal we are working towards. In the second section, I will narrate a grounded story, a real-life story I share with other animals in my surroundings, trying to show how BIO can come to life in a very situated experience in the dense and complex reality we all inhabit.

1. Becoming Into Earthseed

We are not stuck in the inevitable, violent and catastrophic futures for which the apocalypse date keeps being actualized and portrayed on the big screen. We are in the dense present as we keep dragging this fear behind us and pushing it before us. The fear of thinking some apocalypse will inevitably end us, the collective of human beings to whom this prophecy is for. While disregarding the countless lives already lost or damaged by the tentacles of the (always) near end of the world. In this perpetuated freeze mode, we seem to be in, in a world of exceeding information, concepts, articles, theories, media and other contents, while dragged by polarization, radicalization and anti-rights policies, as we are struggling to find our way through extensive workdays and low wages, facing overpopulation, poor health systems; there is an urgency to ask ourselves, what do we wish to become in the face of the future knocking at our door?

Lauren Oya Olamina is the main character in the novel *Parable of the Sower* and *Parable of the Talents*. She lives on a dystopic world not too different from ours nowadays, characterized by heavy migration due to violence and precarity, global warming, profound drug use issues and fascist politics. In this context she writes *The Books of the Living*, the book that contains the philosophy of Earthseed, with which she creates a community to survive in the midst of a catastrophic situation.

The Books of the Living will be our companion through this last chapter, because in it, she shares her philosophy and what she calls her positive obsession, which helped her face the troubled world she lived in. Books of the Living can be thought of as a compendium of her thoughts, feelings and quotes that she has created, either by her experiences or by her readings of various religious texts. I feel a deep emotional bond towards Earthseed philosophy, I go back to these quotes every time I feel powerless and insignificant to the myriads of problems that surround us today. Earthseed's philosophy never fails at making me feel engaged with the world again, which is why I believe it to be valuable and inspiring for the BIO proposal.

I will not explain the story in the novels in detail, I would recommend anyone to read both *Parable of the Sower* and *Parable of the Talents* themselves. But, for the purposes of this chapter, I do have to make two points of clarification about Earthseed philosophy, before I delve into the quotes I have chosen for reflection in this section. The first clarification I would like to make is related to the idea that Earthseed is a religion. Throughout the novel, when Lauren explains Earthseed to others, she is questioned if she is trying to create a new religion. Her answers through the first novel *Parable of the Sower* explain that she is inspired by several religious texts, but she is not trying to build a religion *per se*. In the sequel novel, *Parable of the Talents*, she writes:

> All religions are ultimately cargo cults.
>
> Adherents perform required rituals, follow specific rules, and expect to be supernaturally gifted with desired rewards – long life, honor, wisdom, children, good health, wealth, victory over opponents, immortality after death, any desired rewards.
>
> Earthseed offers its own rewards – room for small groups of people to begin new lives and new ways of life with new opportunities, new wealth, new concepts of wealth, new challenges to grow and to learn and to decide what to become.[2]

This quote shows that, for Lauren, Earthseed has similar elements to those of some religions (especially the ones that exist during her time), which she calls rewards. To her, Earthseed's rewards are different though, from the religions she knows: new ways of life, new opportunities, new concepts of wealth and new challenges. Throughout both novels, she is trying to find these rewards for herself and her community. The rewards become hope for a better tomorrow as she faces a grim present.

Her hope, the hope of Earthseed is bound to a Destiny, clearly stated by Lauren several times during the writing of the Books of the Living. She believes that "[t]he Destiny of Earthseed is to take root among the stars, and there, again, to grow, to learn, and to fly."[3] The second point of clarification revolves around this idea of destiny, because Lauren believes the future of Earthseed is outside of Earth. She believes this planet is a womb, and when human species reach adulthood, they must go out into space. To her, this destiny is fundamental, and she will fight for it as we get to see in the second novel when she embarks on the task of making the space programs succeed.

Destiny is the ultimate goal. To achieve it, Earthseed (when written in small caps, Earthseed refers to all living beings on Earth) must learn to survive on Earth first. Meaning that the Destiny of Earthseed is achieved by learning to live with and in the planet Earthseeds were born in. There is beauty in this destiny, as it is not appealing to escapism, but to radically internal exploration, which can be captured through this quote:

> The destiny of Earthseed
> Is to take root among the stars.
> It is to live and thrive
> On new earths.
> It is to become new beings
> And to consider new questions.
> It is to leap into the heavens
> Again and again.
> It is to explore the vastness
> Of heaven.
> It is to explore the vastness
> Or ourselves.[4]

In this sense, exploring the vastness of ourselves and the heavens (universe) are tasks to perform together, and are the fundamental cornerstones of Earthseed philosophy. Now that I have clarified both points, I will delve into quotes from The Books of the Living in order to weave some of the ideas presented in previous chapters. To do this, I have identified three guiding principles in Earthseed which will work as axis for thought, even though they are all connected in one way or another. The guiding principles are: (1) change is the only constant phenomena to all lives, change is unstoppable, then: if there has to be a God, God is change; (2) in order to keep living, all beings adapt and persist, they shape change, by creating and pursuing their positive obsessions and (3) learning is the most important process we will face as living beings, it is all around us if we are willing to learn.

Principle 1: God is change

Change is not always beneficial, creative, bright and wholesome; it can also bring harm, create havoc and chaos. Lauren knows by experience, change can bear seeds of benefit, as well as seeds of harm. But change can also be shaped by our decisions, actions, relationships and interests:

> God is change
> And hidden within change
> Is surprise, delight,
> Confusion, pain,
> Discovery, loss,
> Opportunity, and growth.
> As always,
> God exists
> To shape
> And be shaped.[5]

Shaping change, for Lauren is a lifelong endeavour, it means to deeply engage and experience the world, learn and accompany one another while we are shaped by change, and as we intentionally shape change too. Everything is always changing, evolving, transforming, becoming, but how do we shape change? What actions do we take to re-shape ourselves and the world?

During the first section of chapter one, I explored the idea of transformation of life through three narratives (*Fledging, Wildseed* and *Oryx and Crake*), in which I explored how transformation of life is directed towards a goal: purity, power and survival. Transforming life is a way to shape change. It can be beneficial, or it can become ill willed. Ina's negation of the transformation of their kind was surrounded by a narrative of purity and hierarchy. Doro's transformation of life was guided by power and domination, opposed to Anyanwu whose guiding principles were family and community. Crake's motives for the creation of a humanoid hybrid to replace human beings were to transform life towards a better possibility of survival.

In other chapters, we can find other guiding principles for transformation of life, for example, in *Barn 8*, chickens are transformed to produce more eggs and meat, in *Dawn*, the Oankali's transform life towards diversity and hybridization, in *Indra's Web*, Mahua shapes life and technology towards a creative and sustainable worldbuilding. All of them are ways to shape change in different directions. Thought like this, these narratives can show us what directions are taking place now in the world in the transformation of life that occurs inside industries, natural reserves, polluted landscapes, liminal places and cities.

We must ask ourselves about this "towards" we direct ourselves while BIO, how to we shape, direct and guide change is something we should be mindful about. What are our motives? Who is being benefited and who is being overlooked? What guides our efforts, are we stuck in the hierarchy, pureness and domination of the Ina's, Doro and the Barn Universe of egg-laying chickens? Or can we move on from it, think of new ways to live, new concepts and new practices to shape life in a beneficial way to everyone involved by creating cities of hope? Can we reshape ourselves enough to rupture the anthropocentric scheme we seem to keep moving towards?

Principle 2: Positive obsession

Having had studied biology, many times I have heard someone saying: extinction is part of evolution, as death is part of life; or some kind of variation of this argument. Although it is true that species eventually evolve and become extinct, I do not think it is the whole story. Not for nothing, the concept of Anthropocene became a huge discussion in academia, and even, many scholars suggested other ways or layers to explore this epoch of human intervention on the planet, proposing the Capitalocene, Plantationocene, Symbiocene and others. I think these efforts are a way of creating necessary accountability, as collective human society, for the devastation we have caused. More so, I think we must not only be accountable, but able to see, grieve and reflect on all the valuable lives that are being lost because of our ways of life, industry, politics and conflicts.

That is why, the Age of Disability became an important part for the third chapter of this book, it hails for accountability but also, it centers the vulnerable by looking for concrete actions of support, mitigation and healing of beings and spaces. To shape change, we must accept to be shaped by change. We must find new ways of caring, we must adapt, connect with others and persist. We can make the healing of living beings and places our positive obsession.

> Prodigy is, at its essence, adaptability
> And persistent, positive obsession.
> Without persistence, what remains
> Is an enthusiasm for the moment.
> Without adaptability, what remains
> May be channeled into destructive fanatism.
> Without positive obsession,
> There's nothing at all.[6]

Positive obsession is not only important for Lauren Olamina, but it was also a fundamental pillar in Octavia E. Butler's life. As she said in several interviews throughout her life, positive obsession was to her the most important driver of her craft. What she characterized as obsession was her love for writing. We could think of positive obsession as what I have called a deep interest, which arises from what Stengers called attachment, a strong motive, an internal urge and an interest that makes one persist on whichever task one embarks on, even when things get complicated. In Lauren's case, when everything she knew, her community and her family is consumed by fire, her positive obsession with Earthseed, makes her hold on to the world and to keep moving towards the future.

In Stengers' idea of attachment, they are the feelings and thoughts that make us act. Both attachments and positive obsession are what make shaping change possible. Because they keep us moving, asking questions, trying to understand and reach out for what we believe. In the paradigmatic cases reflected upon in the third chapter, attachments to conservation efforts make possible many ways of living for the animals cared for. Conservation practices are important and necessary, but they also must be questioned again and again.

When it comes to Northern Rhinos, or any other species being put through intense breeding practices to maintain their species, attachments and positive obsession may guide us through asking questions like: What if we find other ways to confront the loss of this species and the way we treat individuals still living? If there is no mitigation practice to follow, when the systemic problem of illegal traffic is in place; what possibilities of support can we give these animals? How can we offer healing? What would it take for the most vulnerable to survive the conditions we live in? And what kind of life conditions will we work towards for them and others?

In this way, attachments and positive obsession are both anchors and detachments. They keep us on our feet, on our chosen path as we keep moving, questioning and changing. But there are also other elements to positive obsession:

> Kindness eases change.
> Love quiets fear.
> And a sweet and powerful
> Positive obsession
> Blunts pain,
> Diverts rage,
> And engages each of us
> In the greatest,
> The most intense
> Of our chosen struggles.[7]

2. Learning in "sym" Imaginaries

For Lauren Olamina, the learning process we engage in through life is not based on formal or informal education; we learn from everything and everyone around us. If we make learning our chosen struggle, our positive obsession, we open ourselves to the world and to others, as the next quote shows:

> Your teachers
> Are all around you.
> All that you perceive,
> All that you experience,
> All that is given to you,
> All that you love or hate,
> Need or fear
> Will teach you-
> If you will learn.[8]

In Earthseed philosophy, learning must be used to benefit your world, your people and your life. It is a non-ending process of asking questions and seeking answers through different means or senses (touch, smell, taste) and different sources (experiences, past customs, books, people, leaders, thinkers). In the information-saturated world we live in, we might feel already overwhelmed with all the available stimuli to learn with. We must beware of who, where and why we read or search for information and most importantly, we must be able to discern and select information and guide our learning processes.

We can start by engaging in practices of dialogue with others, with books, with all we perceive, which is of course, more easily said than done. But it is important, in Earthseed philosophy, that learning is not a solitary process, if anything, learning is a process we do with others, we learn in partnership, we create learning in sympoiesis:

> Partnership is giving, taking, learning, teaching, offering the greatest possible benefit while doing the least possible harm. Partnership is mutualistic symbiosis. Partnership is life.
>
> Any entity, any process that cannot or should not be resisted or avoided must somehow be partnered. Partner one another. Partner diverse communities. Partner life. Partner any world that is your home. Partner God. Only in partnership can we thrive, grow, change. Only in partnership can we live.[9]

This last quote draws us back to symbiosis and sym-imaginaries, where partnership is what we walk towards. Mutualistic symbiosis can take many shapes, we can create new relationships by this framework, that do not follow the anthropocentric logic of value of different lives and beings. We can shape kin and the world by attentive learning of how to respond better to each other. To do this, sym-imaginaries centre relationships and the learning process captured in them, in an open field where they are able to evolve with time and are not crystallized and controlled processes of utilitarian merchandise.

If anything, BIO is bound to these kinds of learning partnerships that show us new ways to relate to each other with our positive obsession and interests. It is about accepting our possibilities and working towards the renewal of Earth by critiquing the practices that seem unbreakable, by questioning what has been set in stone, and creating possibilities for divergence and change in a world slowly losing any perspective of the future in a positive way.

3. BIO: The Attentive Co-shaping of Kin and World

> There is no end to what a living world will demand of you.
>
> —Octavia E. Butler, Parable of the Sower

This is a grounded story involving humans, cats, squirrels, *tlacuaches* (opossums), *cacomixtles* (ringtail cats), arachnids and many other forest ravine inhabitants. It is a story I know too well, because it is a story of becoming into others that me and "J" have shared with other inhabitants surrounding what we call our home. A little backstory is needed. I moved into this house ten years ago, in a neighbourhood that has since grown unprecedentedly because it is near Santa Fe, a place that used to be a trash ground and is now full of office buildings, malls and skyscrapers. Santa Fe was connected more than ten years ago through three huge bridges built over ravines to San Bartolo Ameyalco, Centenario and Las Aguilas neighbourhoods. Being the latter, the one we live in.

Since then, the forest and natural reserves that existed have become housing complexes, some of them not even half full, but that have taken over most of the forest where many animals and plants used to thrive. When I first came to live here, it was common to see many arachnids, spiders and scorpions, to hear owls at night and to see opossums walking calmly in the garden. As liminal animals whose home was being swallowed by housing complexes, they used the spaces available, they moved and adapted.

Not only did these animals face the reduction of their natural habitat and the pollution of the nearby river; they also faced the reproduction of feral cats. Even now, people tend to let the cats roam free without caution. When kittens are born though, they are abandoned, this has created a big community of feral cats. Feral cats are especially damaging to an ecosystem, even more so an ecosystem already invaded by less furry but still hostile human beings. In a sense, human invasion on the ravines was expanded by their irresponsible practices with domestic cats.

When cats started annoying humans, fighting for territory, allegedly scratching cars and entering their houses; humans called the local animal protection agency. They resorted to capture, but spaces in animal sanctuaries in the city are minimal, so they usually euthanize them. People did not care, as long as the cats were out of their way, for them, it was either that or using rat poison mixed with cat food to get rid of them.

This absolute disregard for the cats' lives ended my relationship with a cat I knew for years, the mother of many kittens I cared for, some of whom had to be set up for adoption and sterilized; others who, sterilized, still to this day live freely in the garden. Pelucas was taken from us thinking no stray cat would be missed. She is greatly missed. Many cats remain though, some come here for food and rest, some we only know by sight and random cuddles in the streets. But there is no campaign or care practices for cats or for any of the ravine's original folk.

This story is a story of BIO, a story of deep interest and care for not only the cats, but also the place we live in and share with others, including humans. It has been a story of learning through response-ability what we can do to make their lives better and for them to have a safe place. There are many discussions possible in the midst of this story, and I do not have many answers to debates that can be linked to it, but hopefully sharing my train of thought, can help see how a BIO framework can aid us to achieve not a recipe for success, but the development of a positive obsession in learning to live better with others.

a. Cats vs the ecosystem

There is a non-intervention debate in ecology and environmental studies that considers that humans should not intervene in natural places and relationships. This principle got important recognition in many areas, being now mainly a state's concern, which is why there are territories marked as Natural Reserves, Natural Sanctuaries and other forms of protection of natural spaces. But this principle only includes natural landscapes, places mainly untouched by human expansion. Liminal places, spaces where the remaining ecosystem lingers even as it has been profoundly intervened by human activities, are not considered natural and are up for the taking. What should we do, then, with liminal places, with the patches of ecosystems that remain, especially as human settlements grow and take over, and there are fewer natural places and more liminal ones?

Then, there were the cats, representing another material danger to many inhabitants of the ravines (arachnids, insects, small reptiles, mice, squirrels, opossums and ring-tailed cats). Cats are not responsible for the ecosystemic collapse and are, also, valuable and important living beings. It is overwhelming how easily humans use the term pests to justify bringing upon them death. Feral cats are considered a danger to the ecosystem, they are invader species and as such there is a scientific and ecological justification that approves and legalizes their massive assassination.

In this debate, I always feel I am on a conundrum. As a biologist, I understand that invader species are a problem to the ecosystem and the beings who live in it. And I know many people on field recognize the problem is the introduction of these animals to an ecosystem which is, in most cases, made by human beings. But as an animal studies enthusiast, I am against the classification of value in which some are allowed to live while others are being disregarded. Invader species are still sentient and aware individuals and families, who have no more fault for the ecosystemic collapse than the humans who introduced them.

This paradox has created an idea that we cannot value, care for or defend both the ecosystem and the living beings we have classified as invader species. In animal studies that has been thought as the impossibility of thinking with both environmental and animal studies, because of what Barbara Noske describes as the individual versus ecosystemic reductionisms.

Noske analysed that animal studies entered an individualistic reductionism realm, in which they concentrated on the individual animals, their sentience, value and necessary legal protection, especially at the industry level or with domestic animals. The goal is mostly animal liberation or animal welfare. On the other hand, environmental studies have concentrated on animals only as far as they are part of the ecosystem, they are concerned with the whole community of beings in natural places, falling into an ecosystemic reductionism realm. So, she asks:

> How are we to navigate between individualized ethics and ecosystemic reductionism? The animal lobby bestows on the sentient in nature a status of individual humanness: it asks how animals are part of human society and ethics. The movement could perhaps bridge the gap which separates from deep ecology by overcoming its exclusive focus on sentience. It could extend its compassionate ethics so as to include the non-sentient and even the inorganic. The tricky part would be how to include the whole earth without simultaneously humanizing and colonizing it. Moreover, there always will be clashes of interest between animals and animals, animals and plants, individuals and species, the organic and inorganic.[10]

Let us be clear that when Noske was talking about these reductionisms, more than twenty years ago, critical animal studies were on their initial moments and multi-species studies were still on the way. Nowadays, there are many possibilities to think outside these reductionisms, but they have not been entirely overcome. Scholars like Catia Faria and Eze Paez propose that animal and environmental ethics are completely incompatible: "[o]ne can endorse one of the two views, but not both at the same time. One reason why animal ethics and environmental ethics are incompatible is because of the different criteria of moral considerability they endorse. Another reason is because of their different normative implications."[11]

The criteria for considerability in animal studies, according to Faria and Paez, is sentience and this criterion is not mentioned in environmental studies. Indeed, many environmental ethics barely consider suffering of animals in nature, as most situations are considered natural and outside the moral spectrum (e.g. a lion killing a zebra) and because, there is a non-intervention consideration. Sentience is a spectrum studied by animal scholars, and it is mostly applied to animals inside industries or in close relation to humans, like domestic animals. This creates a situation in which environmental studies have overlooked the suffering of animals in the wild and neglected to care for them:

> In sum, the majority of sentient individuals living in nature plausibly have existences containing more suffering than positive experiences. Because they matter, we have reasons to prevent or alleviate the harms they suffer whenever we can as well as find new and better ways to help them. Since there are so many wild animals and their lives are so miserable, these reasons are very strong. Because environmentalism cannot accommodate this important commitment, it ought to be rejected.[12]

The incompatibility between these two frameworks has created polar studies and fields. Faria and Paez believe that speciesism studies can dialogue with environmental studies, but in reality, those conversations do not last long. I have also experienced this directly, by studying biology, I knew and felt the reticence that many people in environmental studies have for people from animal studies, whom they generally portray as activists and not serious scholars. In a way, since graduate school I have been on both sides of the coin and heard arguments from both sides.

Although I have no absolute answer to this debate, I believe a BIO framework may help us situate ourselves in the middle of the discussion, recognizing we do need a framework to work with these liminal situations. Or as Haraway would call it, to think in the muddle, to ask new questions, to explore them again and again. For this, I will call back to Taylor's disabled ecologies framework. There are questions we can ask about this situation in a more propositional way. The cats' vs the ecosystem or the invader species vs the local community dilemma does not need to end on a discussion of whose ethics has the best criteria, there are more pressing issues on this dilemma than the configuration of an ethical hegemony.

So, first, in terms of mitigation, we may ask: Who is accountable for the root problem? Who is accountable for the liminality of ecosystems? Who is responsible for letting loose a species who might become a problem to a habitat? How do we find ways to mitigate the problem? Who should we talk to, organize and work with?

Second, in the healing and supporting realm, we may ask: Is there a way to value both, the renewal of a disabled ecosystem and the vulnerable beings who lost their home, and at the same time, those introduced to the ecosystem and now may be causing problems? How can we work on healing the ecosystems by holding space for animals and natural spaces to flourish? Can we help invader species have a good life while still trying to keep balance in the ecosystem and with other animals in it?

Now, that all sounds good and dandy. But let us go back to the story. I was concerned about how to find balance between a cat population on the rise and to hold space for other animals of the ravine. The first principle I stood by was not to damage any animal, and to try, with my own limitations, to minimize suffering in any of their lives. This is not always easy, and of course, not always in my control. Sometimes suffering happens, and you cannot do much about it. Sometimes, when cats get into fights and get seriously injured, they hide somewhere inaccessible, and you are not able to help. Only once did an injured cat come to the garden door crying and showing me his damaged paw, most of the time they do not. In a sense, this is why, as Haraway suggests, there is an art to living a dying with others. It is unending learning to try to find the best way to respond.

Anyhow, I was faced with a dilemma concerning the debate between animal versus ecosystem studies: should I feed the cats? I thought, on one hand, feeding them would reduce, at least in a tiny way, their hunting of other animals; on the other hand, feeding cats involves the meat industry. Cats are strict carnivores, they need meat. Unlike dogs who are omnivorous and with a very special diet might be perfectly fine eating a vegetarian diet. This is a big dilemma, as I believe the meat industry is one of the most abrasive industries existing today, and I am always trying to avoid participating in it as much as I can, by being mindful about what I consume.

Finally, I decided to feed the cats. Which means I do participate in the meat industry. There are many reasons to do so. It is a way from keeping them from wandering territory looking for food, it is a way to avoid some fights and some hunting. Also, feeding them creates some trust even as they still are wild and feral. Feeding them might allow me to capture them for sterilization or if they fall sick, it also allows me to give them medication for parasites that could be lethal not only to them but to other animals in the environment, or medication for injuries they might have.

Feeding the cats is nowhere near a perfect choice. It is frustrating that we humans have intervened so much at this point that the way to feed invasive species like cats, insert us in the industrial meat complex, or if we want to avoid it we must start producing animals to feed them. We are all in the muddled stories of the world we have been inherited with, and we, whether we like it or not, learn to live with what Haraway calls the scandalous thing (power laden white capitalistic patriarchy).

That is why I also hoped, by care, sterilization and the adoption of some of the cats I could reduce their population (without resorting to poison or euthanasia) and giving them a chance to have a better life. And by reducing the population, I could also reduce the amount of food I got for them. I have done for at least eight years. I cannot say, after all this time I have a population study to back up that this indeed happened, that cats have reduced their numbers by this method, and that they do not prey on other animals as much.

As I said, there have been several instances in which neighbours use deathly methods to deal with the population of cats. What I have tried to do is find ways to be less deadly, to find alternative long-term ways to deal with the problem of the introduction of cats in a liminal space, focusing on mitigating, healing and supporting them, as well as other animals I have encountered. There have been times in which I have been able to rescue baby opossums from some of the cats, this is not a common occurrence and, also, it has been a long time since I have seen a carcass of a bird, reptile or mammal around the premises of the garden.

The reality of relationships is always complex, in a sym-imaginaries way, what I can do is learn and inform myself about the options available, make difficult choices, ask questions over and over, to explore the vastness of the partnerships I have encountered and cherished. I can be attentive and alert, ready to help any animal I come across and to care for the ones that have made this place their home. I cannot control what people around me decide to do, I cannot control what cats do, I can only guide my own actions by a positive obsession of learning to live and respond to others. I have limitations of my own and I am limited by the world I live in.

I cannot avoid all suffering, but I can at least try to avoid some suffering sometimes and try to project a future in which we are able to better care for animals in the wild, inside the industrial complexes (or better yet, without the industrial complex) and in the liminal spaces we inhabit with them. In a sym-imaginaries way, the relationships we encounter matter, creating community matters, we create associations for better or for worse. And, hopefully, we do what we can to partner life. Or in Olamina's words: "Partner one another. Partner diverse communities. Partner life. Partner any world that is your home."[13]

b. Cats' adoption and sterilization biopower

Here is another dilemma I faced while learning to respond better to the cats around the neighbourhood. Like the Oankali did to Lilith, I took a decision for them about their bodies. I exercised biopower practices of control over their bodies and populations. I sterilized them. I am aware that these are actions that take the animals' bodily autonomies and their freedom of deciding their own future and who they want to spend it with. Maybe we do not usually think of it that way, in many animal activist circles neutering animals is portrayed as an act of kindness. Sterilization lowers the suffering of next-generation animals, because they will not exist in such big numbers.

Maybe we do not question this too much, because humans have the legal right to decide for a cat or any feral animal body. That is how the system is configured, people can legally own cats and decide for them if they will reproduce or not and even do business with their reproductive capacities. In an animal study perspective, it would probably not be well seen to do so. In an ecological perspective, the use of reproductive capacities is wildly normalized for conservation efforts, and no questions are raised about the animals' bodily autonomy. Some might say this is because autonomy is not a concept that can apply to animals, and maybe this is true in the way it is configured in the legal sense. Let me be blunt, bodily autonomy is not even recognized for many humans in the world we live in.

Industries and conservation programs aside, as I have said before, feral cats reproducing in a liminal area can cause problems to the already damaged ecosystem and the animals living in them. And as I have also said before, other problems will arise, like humans trying to contain the population with deathly methods. Then, the exercise of biopower that we embark on while sterilizing cats, seems the lesser of the bad possible situations. But is it in the best interests for individual cats? To answer this question, I had to learn first what reproduction means to cats, their bodies and their social dynamics.

I must say, this question must be considered on a case-by-case basis, in this case, I am talking specifically about what I have learned from my association with Pelucas (female) and Félix (male). There are also some facts I researched and others that my veterinarian shared with me. These are what I have learned: For male cats' reproduction is one of the main causes of aggression. Male cats will fight for territory and mates. The social dynamics are mostly stable while a male is mating with a group of females, until another male comes along to fight. I've seen many cats perish from fatal wounds or isolation after these kinds of events.

Félix, who is now a senior, has been in many fights and fortunately, he has made it out alive, but not without a scratch and many times even much more than just a scratch. He has had his paw broken, his head opened up by bites, his neck wounded to the point of heavy inflammation, and many other difficult situations. He is absolutely feral, which means he has never accepted any help, he keeps his distance and is pretty distrusting, so I have never been able to capture him for sterilization. All I could do was to hold space for him as he fought for territory and females.

For female cats' reproduction is even more costly. Female cats may get pregnant every 6–8 months, gestation lasting 60–65 days, they may be pregnant again before weaning off their babies. They may have between 2 and 8 kittens every time. These kittens will be ready to reproduce as early as four months. If you do the math, this is a whole lot of kittens. Reproduction for any mammal is costly, their bodies go through many hormonal and morphological changes and use much energy, either for what they consume and their reserves. And caring for their offspring will take away from them even more energy. Some mom cats hunt and feed their babies first, sometimes even to the point of malnutrition. Many cats after several births will experience gestation problems, prolapsed uterus, dead foetus or sepsis.

Pelucas was one of Felix's main mates. She came along after another female cat (who I called Juanis) left the premises after leaving her offspring in the garden. Pelucas became a regular because she was soon pregnant and needed food and space. Her first litter was born on the other side of our garden premises, which is private property no one, except cats, use. I never got access to the babies as Pelucas was always weary. Her babies were still not weaned when she was pregnant again.

The day we finally caught Pelucas to take her to sterilization, after three litters (from what I know), she was already pregnant again. The veterinarian told us this news as he made an ultrasound on her belly, he straightforwardly said we had to make a tough decision. She could be sterilized but the babies would be of course aborted, or she could go back home and give birth. He assured us though, it would not be a safe birth, she had a thin uterine wall by then, probably because she had been pregnant several times before. She could die from a uterus tearing before or during giving birth and the babies would surely die too. With a heavy heart, we decided to sterilize her.

From an outside perspective looking in, this could be set in a sense of classical animal ethics principle, the less suffering for the less animals. If Pelucas gave birth the possibility of all of them dying was great, if we sterilized her, at least she would survive. And mind you, she already had three kittens waiting for her back home. It was a hard decision, I felt as though I betrayed her, I still wonder if we made the right choice.

I go back to Olamina's Earthseed verses to remember that hidden in change there is: "surprise, delight, confusion, pain, discovery, loss, opportunity, and growth."[14] Change shapes and it is shaped by our decisions, our passions, our loves and hopes; it is shaped by our anger, frustration, ignorance and sadness. In the end, change remains the only constant, suffering exists and expands, partnerships thrive and break, living beings are pushed into extinction, and transformation is expanded on a one-way street to human perfection or immortality.

But we are still here, still alive, still able to change. We can still ask questions over and over, we can still learn better ways to respond, to care, to support. And as Lauren Olamina taught me, to benefit our world and our partnerships, we must consider consequences, minimize harm, ask questions, seek answers and learn,[15] over and over, what we could do to be less deadly, kinder and more supportive to all who surround us.

Pelucas showed me many things about love, care, loss and trust. She was a teacher to me and to her kittens and we were willing to learn. She will always be missed and remembered by us, the animals living in this piece of world, this piece of ravine we call home. Animals who are trying and sometimes failing, to create attentive ways to co-shape kin and world, trying to make ourselves partners to life by BIO each day together.

We give our dead
To the orchards
And the groves.
We give our dead
To Life.[16]

Notes

1 Octavia E. Butler, *Parable of the Sower* (London: Headline, 2019), 3.
2 Octavia E. Butler, *Parable of the Talents* (London: Headline, 2019), 311.
3 Ibidem.
4 Ibid., 265.
5 Ibid., 89.
6 Butler, *Parable of the Sower,* 2.
7 Butler, *Parable of the Talents,* 43.
8 Butler, *Parable of the Sower,* 265.
9 Butler, *Parable of the Talents,* 129.
10 Barbara Noske, "Two Movements and Human-Animal Continuity: Positions, Assumptions, Contradictions", *Animal Liberation Philosophy and Policy Journal* 2, no. 1 (2004): 8.
11 Catia Faria and Eze Paez, "It's Splitsville: Why Animal Ethics and Environmental Ethics are Incompatible". *American Behavioral Scientist* (2019): 4–5.
12 Ibid., 11.
13 Butler, *Parable of the Talents,* 129.
14 Ibid., 89.
15 Ibid., 59.
16 Ibid., 5.

Chapter 8

BECOMING-INTO-OTHERS: A SYMBIOPOETIC APPROACH TO RECONNECTION WITH NATURE

> Should not the whole theory of the subject be reformulated, seeing that knowledge, rather than opening onto the truth of the world, is deeply rooted in the "errors" of life?
>
> —Michel Foucault, *Life: Experience and Science*

> … a living form can be interpreted as an organic consciousness that is able to maintain its form and to transform this form by incorporating otherness.
>
> —Audronė Žukauskaitė, *Organism-Oriented Ontology*

> If one clearly identifies this source of human emergence into history, one can also understand that man is more holistic than nature. His dream of a universal rights-based commensality is the best thing that earth has produced.
>
> —Peter Sloterdijk, *Nature to-be-Made. The Crucial Subject of Modern Times*

So far, the three chapters reviewed on three different effects of biopower in recent times (transformation, extinction and symbiosis) have shown some kind of paradox: how is it possible than in the age of biopower in which transformation reaches different levels or realms of life (from genetics to symbiosis) it is also the time in which humanity is facing the massive extinction of species? In other words, how is it possible that human beings are focused on transforming life while many species are near to extinction? In a biological perspective could be answered these questions taking as an argument the main idea that life while changing through time eventually will face extinction. If life had a beginning, then life must have an ending in certain times. In an evolutionary level, species emerge and might become extinct reaching certain unfavorable conditions, such as food shortage, environmental change, epidemic diseases, the invasion of another species, floods, droughts or even a meteorite. Then life and species are subjected to a certain process of extinction. Through this natural perspective, the notion of extinction is a natural process.

It seems that life, by transforming itself, puts itself at risk of extinction. It is not just about life trying to sustain itself at risk of not surviving. Thus, the feat is not reproducing but surviving and avoiding extinction. In this perspective, life is always on the verge of extinction. Facing death and extinction, life had to learn how to reproduce. So, extinction is not an exterior condition to life. In fact, it is the core from which life proliferates. In this case, life and death/ life and extinction are phenomena considered in the natural process. Within this relation, every living being is capable of becoming into others across time. In Simmel's words, life transcends itself, because life is always more-life. But considering the spiritual realm in which human beings transcend the boundaries fixed by life, then the natural process has been transformed by human hands. In doing so, the relations between life and death and life and extinction have modified their natural dynamics.

Now, facing global warming and a series of catastrophic events, such as the sixth massive extinction, the biological discourse of symbiosis has become the new answer for playing within the paradox: in order to avoid extinction human practices must transform, in the first place, their relationship with other organisms and, in the second place, the relation among organisms (and technorganisms). However, it is necessary to think that the entry of symbiosis in biohistorical and biopolitical domains is itself a denunciation of the process of disconnection that human beings have achieved with nature in general and with other organisms in particular. Even on the planetary scale, there has been a great disconnection, which is evident, at least, in the garbage dumps that proliferate in certain places on the continents (such as Apex Regional Landfill, located in Clark Country, Nevada, USA). These sorts of places are evidence that human beings do not care about the planet, or they do not think about the consequences their waste produces. Thinking that garbage and waste disappear from their field of vision, human beings fail to make the connection between their actions and environmental damage. In this case, symbiosis understood as discourse and practice encourages reconnecting humans with other organisms, environment and even with the planet (Gaia). Considering the current situation, it does not mean that human beings must control symbiotic processes, in biopolitical terms, nor that they just mimic symbiotic relationships. In the era of symbiotic turn, human beings could mimic the power *of* life that lies in symbiosis: different forms of association that, in the best cases, produce benefit for organisms involved. At the same time, it would be a matter of avoiding parasitic relationships. Imitating the power of symbiosis opens up the opportunity for organisms to become-into-others. Within this framework, this chapter aims to design a symbiopoetic BIO, that is focused on transforming the interference between life processes and historical movements according to the symbiotic power of life. This approach must consider both biological and technical symbiosis in order to think of a new way to reconnect human beings with other organisms, environments and planet Earth. To meet the objective, this chapter is divided in the following sections: *Avoiding terrors in Earth* relates the problem with plastics to a sf story to describe the monstrous effects of technological products and the technosymbiotic human condition; *From technical disconnection to reconnection with techno-nature* examines the historical conditions by which humans have disconnected from nature and the technical reconnection possibility; *The technospheric hyperconnection* explains four cases in which the connection, disconnection and reconnection of nature and humans are shown; and *Towards a symbiopoetic reconnection with nature* determines the conditions under which humans can imitate the power of life, considering the life-giving power of trees, to reconnect with nature.

1. Avoiding Terrors in Earth

Technique and technology have become a type of symbiont since the beginning of human history. In some cases, symbiosis could have been mutualistic, commensalistic or parasitic. In recent times, after the Industrial Revolution and the excesses of the twentieth century, waste from technical objects has turned into a parasite. Proof of this is the huge amount of plastic and microplastic now found in the deep seas and in the stomachs of certain birds and marine animals. In the same way, the function of photosynthesis in plants is being affected by what is now recognized as plastic stress. And in humans and other animals, plastics and other chemical substances are disrupting endocrine functions by interfering with hormone receptors. Probably, after nuclear and climate terrors, the fear of plastics is shaping new ways of relating to technical objects: from recycling to rejection campaigns. But despite all the efforts, plastics continue to flood the planet. In a wide range, passing from micro- to macro-scale, parasitic technology is now circling Earth as space junk.[1] So, in this case, the famous Technosphere, named by Peter Haff,[2] is a sphere of waste and garbage. Considering this case, since the second half of the twentieth century, humanity has witnessed not only the proliferation of technical objects, but also their waste and garbage, which fail to be incorporated into the ecosystem flows for recycling. In the most catastrophic future, waste and garbage will bury the biosphere. However, facing that reality, solutions are emerging, such as the use of different organisms, fungi and bacteria, capable of *eating*, in this case, plastic. So, human beings cannot stop producing plastics, but they can instrumentalize, as an exercise of biopower, different life forms or living beings to control their own waste. To avoid the parasitic relation with plastic (the ultimate technosymbiont), human beings are establishing mutualistic associations with fungi and bacteria. With this strategy, the hope is to address the major environmental problem generated by plastics.

But as with every strategy, it could have a counter effect. It is necessary to think about the consequences of using living beings as biotechnologies. Understood as a warning, Anatoly Dneprov's[3] story titled *Crabs on the Island*, published in 1958, tells an experiment which seeks to "test Darwin's theory."[4] Despite this theory is related to biology, natural selection and evolution, nothing stops Dneprov to imagine that a "most strange-looking apparatus"[5] comes to life. Besides having "six large articulated claws, it had in front two pairs of slender tentacles whose ends were tucked into the gaping protuberant 'mouth' of the hideous best. In a depression on its back gleamed a small parabolic mirror a dark-red crystal in its center. Unlike an ordinary toy crab, this one had two pairs of eyes, in front and behind."[6] This particular apparatus only needs solar light, water, sand and certain metals not only to survive, but to reproduce itself. After passing certain time, when there are already too many crabs and few resources, the experiment begins, that is to say that begins the primordial scene of Darwinism: the struggle for life. Crabs do not only fight for food but destroy and eat their opponents. In doing so, the winning crab reproduces itself giving birth to an enhanced crab: bigger, stronger. The aim of the experiment is to improve the robots through struggle, because in the end only "the most perfect ones will survive."[7] So, every new generation, rather than being a mere copy of the previous one, reproduces a different mechanism "that bears no resemblance at all to the original."[8] Every new copy is better than the other. The experiment is successful, because improved crabs appear, but they end up hunting and killing the creator for carrying the last pieces of metal in his dental prosthetics. When the creator believed that his creation could become a great weapon in times of war, because the crabs could only eat all the metal from the enemy's weapons, never thought that he would die by the crab's claws. With this experience, the remaining character of the experiment learned a moral:

> I thought of how many able people in our days had used the powers of reason to do harm to others. Cookling's invention, for instances. I was sure it could have been used to good purpose in metal mining, possibly. The evolution of these creatures could have been so directed that they might have performed that function with the utmost proficiency. I came

> to the conclusion that if the machine had been perfected properly it would not have degenerated into a gigantic clumsy monster."[9]

Just like many other dystopic stories, the main problem observed by Dneprov is not the use or misuse of technology but the purposes it aims at. It could even be said that the real problem with the experiment is not the misuse of the technology (living technocrabs), but rather its potential side effects, discrepancies, or unanticipated variables. This sf story that imagines a type of technosymbiont anticipates the use of living organisms that, while they may have good purposes, can also become a threat or a possible risk not only for human life, but for life itself. Therefore, in a time of symbiosis and technosymbionts, it is important to pay attention to the associations that are established with both technical and biological symbionts. Just as technocrabs, plastics are threatening life. And the problem with plastics is not their purpose, neither the use or misuse, but the side effects, discrepancies or unanticipated variables. Plastics have reproduced so fast in the last decades that, just like the end of Dneprov's story, they have become one giant monster, whose magnitude, if it continues to grow, could eventually exceed the magnitude of the biosphere itself.[10]

Contrary to the end of Dneprov's story, recent human beings cannot leave behind technique, technology and Technosphere, because, as Oliver López-Corona and Gustavo Magallanes-Guijón, they have become technobionts[11] which is the effective realization of the link between human beings and technology in a co-evolutionary sense. That is to say, human beings do not only establish multispecies relations, just like holobionts, but also multi-technique associations. For this link, human beings must consider technology in a deeper sense to make other kinds of decisions about technological uses. López-Corona and Magallanes-Guijón propose that, in the Technocene understood as the Earthly time in which technology became a geological force, human beings still could be *Anthropos*, but can cease to be technobionts in the following sense:

> …We sure cannot stop being *Anthropos* but we may certainly stop being Technobionts. A word of warning here, by no means we are proposing to neglect scientific or technological progress, nor we are thinking we should live as hunter-gatherers. We are merely saying that we need to take technological coupling into account when trying to understand Earth System Dynamics and that some types and intensities of technological coupling should be treated with the maximum application of the non-naïve (this is key for not falling into misunderstanding) Precautionary Principle.[12]

This explanation considers what Dneprov's story highlighted: that technology development must consider the consequences in the first place. Purposes could be acceptable and beneficial, but side effects could be catastrophic. In another ontological dimension, what is at stake in these explanations is the disconnection of human beings and technology (or technobionts) with nature or, in new technical words, Earth System Dynamics. And in terms of biopower, these explanations must shift the biological paradigm in order to achieve different consequences. In this case, the biopower experiment needs to pass from Darwinian theory to Symbiotic approach. This new biopower experiment must take into account, in first place, different relations and associations between biological and technical symbionts, considering the purpose, consequences and side effects and, in second place, a new way of reconnecting human beings and technology with nature, insofar as it produces a different kind of interference that transforms life process into a historical movement.

2. From Technical Disconnection to Reconnection with Techno-Nature

Now that contemporary human beings can understand their own existence as technobionts inhabiting the Technosphere in the epoch of the Technocene, it will be easier for them to realize their disconnection from nature. In this sense, the only important point would not be having the mental clarity to return to nature, as long as human beings stop using technology. It is as if contemporary human beings, in order to move away from the chaotic image generated by the Anthropocene, identified or assumed an ecosystem function, the benefits of which would only be achieved without technical mediation. After all, human beings can continue to be considered biological singularities that, like other species, perform an ecosystem service, just as forests regulate climate and absorb carbon dioxide. Thus, it would seem that the important thing would be to put aside technology to perform the ecosystem service or function. For example, in this way of thinking, the emblematic figure of Henry David Thoreau stands out:

> Here is this vast, savage, howling mother of ours, Nature, lying all around, with such beauty, and such affection for her children, as the leopard; and yet we are so early weaned from her breast to society, to that culture which is exclusively an interaction of man on man – a sort of breeding in and in, which produces at most a merely English nobility, a civilization destined to have a speedy limit.

> In society, in the best institutions of men, it is easy to detect a certain precocity. When we should still be growing children, we are already little men. Give me a culture which imports much muck from the meadows, and deepens the soil – not that which trusts to heating manures, and improved implements and modes of culture only![13]

Thus, advocating for a return to nature would have to begin with the question: if they return to nature, which in this sense would mean leaving culture (and Technosphere), would human beings, in and by their very biological uniqueness, fulfill an ecosystem function? The question is not in vain if one takes into account that at the same time it affirms that, indeed, human beings experience a disconnection from nature that occurs, as Thoreau asserts, through an early weaning that could not only be recorded in the generational birth of human beings who inhabit contemporary super-technological societies, but also in anthropogenetic terms that would define the point of secession at which certain human groups, due to both their biological and technical evolution, gradually separated themselves from nature. Thus, the early weaning pointed out by Thoreau could well be the ultra-historical moment that insistently repeats itself in human history since it occurred or originated in its prehistory. Thoreau's culture[techno]phobic approach, in addition to denouncing the collapse of human civilization (*to have a speedy limit*), is a historical effect of, as Sloterdijk says, "the battle between an archaic couple that has catastrophically split up after an original symbiosis [Nature-Human beings]."[14] However, the original split is what has produced human beings. In Sloterdijk's words:

> From the viewpoint of historical anthropology, man's becoming human was, from the first, an effect if his progressive release from his surrounding environment. People became people because they were able at an early stage of their evolution to escape the absolutism of nature, seen here as an environmental pressure, and to make themselves at home in a niche completely filled with group alertness, readiness to flee, mobility, empathy and cooperation – the qualities, more or less, which modern specialist consultants see as the optimal management and teamwork strategies.[15]

From this historical situation arises an assumption in terms of philosophical anthropology that could be stated as follows: human beings, being disconnected from nature, are those strange animals that seek to connect with it in other technical ways. Thus, the technological dimension cannot be denied. Consequently, if technology (or culture) once disconnected humans from nature, it could also reconnect them.

3. The Technospheric Hyperconnection

To appreciate the scale of the new ways in which human beings reconnect themselves through technology with nature, the existence of Voyager 1 can be considered. This robotic space probe was launched on September 5, 1977. To date, it is the furthest object from its point of origin and its creators. In its lonely journey, if it were capable of experiencing emotions, it might well encounter the same sense of terror that humans discovered in the modern era: the realization that the universe is vast and seemingly empty. However, this technological artifact has traveled such a great distance that no human being, currently, can do the same, lacking both the technological and physical capabilities. Voyager 1, now in interstellar space, continues to transmit data about the cosmos. Thanks to this spacecraft, humankind has gained access to knowledge about parts of the universe that would otherwise remain a mystery. But if we consider the use of such objects for exploration, we must also acknowledge the existence of satellites for Earth observation, communication, navigation, meteorology, space telescopes and space stations. However, this generation of knowledge comes at a high cost, as it has resulted in space debris that now surrounds the Earth. In this sense, the relationship between satellites and space debris represents the most extreme example of technological impact, leading to the concept of the Technosphere, which is "[…] a geologically young global system, is Earth's newest sphere, joining the four classical spheres of air (atmosphere), water (hydrosphere), rock (lithosphere) and life (biosphere)."[16]

From these examples, it becomes clear that technological dimensions have already shaped, in some way, the environment that now surrounds the Earth. Recognizing this technological overlay, the perspective of what is normally seen as the iconic image of planet Earth, represented as the "beautiful blue marble," shifts to an image that foregrounds all the artificial satellites and debris, which, orbiting the planet, appear as a kind of superimposed ornamentation. Furthermore, to these two images – one showing the planet in its natural state and the other showing the planet adorned (from cosmic aesthetics to planetary ornamentation) – we must add yet another image, produced by the data collected by satellites monitoring global warming. This perspective of the globe changes once again: from a planet highlighted by its blue color, the image shifts to yellow, orange, or red due to rising temperatures. These thermal images provide evidence, on the one hand, of the technological layer that now covers the planet and on the other, of global warming.[17] Consequently, the issue of technology, from the array of satellites and space debris surrounding the planet to Voyager 1 traveling through interstellar space, has transcended not only certain planetary boundaries but also the limits of human capabilities: who will be able to retrieve Voyager 1 if it malfunctions? However, thanks to these artifacts, a new form of planetary hyperconnectivity has emerged, shaping the Technosphere in other dimensions.

This brief commentary on satellites and space debris serves to reflect on the reality of the Technosphere. It is not intended to deny the technological dimension, nor to portray everything as inherently bad and destructive, nor to claim that everything is good and positive. Rather, the aim is not to simply choose between technophobia and technophilia, but to broaden the scope of analysis, which in this case can be linked to the exercise of power over life, or biopower. In this sense, the proposal is to analyse the exercise of power over life as it relates to technology and biotechnology, given that, by intensifying, expanding and deepening this exercise of power, these technologies transform life itself.

The ways in which biopower is exercised, tasked with completely invading life in order to transform, control, produce and foster it, by taking both the body as machine and the body as species as its object, imply a link between behaviors, gestures and movements with the control that involves reproduction, birth rate, mortality rate and the control or regulation of endemic, epidemic and now pandemic diseases. In short, it can be said that to analyse the exercise of power over life, one must first consider the poles of the anatomopolitics of the body and the biopolitics of the population, which in turn would mean showing how the body-machine and the body-species are produced.

Several criticisms have been levelled against Foucault's historical analyses of biopower, among which the most prominent is the accusation of his anthropocentric bias, for allegedly focusing only on human beings when studying the subject and power. However, rather than simply criticizing or accusing him of anthropocentrism, one should use the conceptual tools that Foucault bequeathed to future generations to analyse the exercise of power in a different way, one that, by transcending the human realm, produces other types of effects on life. There is now much literature on this subject, but among the first approaches, the book entitled *Discourses of the Environment*, edited by Éric Darier, is noteworthy, as it suggests that the exercise of biopower is not limited to human populations, but can also be directed at the environment. Thus, the exercise of power captures, or attempts to capture, all life in general, as Paul Rutherford affirms, referring to Foucault, that if life enters history, it is not only human life, but all life, or life in general. Therefore, environmental policies that seek to control and regulate nature can be analysed as a form of regulatory control. Other analyses, such as those by Dawn Coppin and Cary Wolfe, have sought to demonstrate that the exercise of biopower, in its anatomopolitical and biopolitical dimensions, is also present in both industrial farms and experimental laboratories where animals are bred and used, whether as machine-like bodies that produce milk or meat, or as bodies whose biological characteristics are manipulated for drug testing. Another important book on this topic is Dinesh Wadiwel's *The War Against Animals*, whose analysis also addresses the disciplinary practices applied to animal bodies in industrial farms and the biopolitical controls over their biological processes. These processes involve not only reproduction, but also the need to increase their size and weight. Since the industry needs to produce more meat, vaccines and other medications are administered to prevent disease and growth hormones are used to increase production in a shorter time. In industrial animal farming, there is both disciplinary control over the animals' bodies and biopolitical control over their biological processes, just as the symbiotic processes occurring in cows' stomachs are now controlled to reduce methane emissions. Thus, the exercise of power completely permeates life, whether that of animals in a nature reserve or that of animals raised on an industrial farm.

Now, when distinguishing between the management of life in general and the biopolitical management of both human beings and other animals, it is important to specify how to analyse specific cases. These could be analysed individually to highlight their particular significance, but it is also possible to adopt an analytical approach that reveals the indistinction of the exercise of power over life. Just as animals are produced in industrial farms – a way of producing bodies and, in a sense, life itself – there are other types of biopolitical practices directed at human beings, such as organ transplants, which involve the biopolitical management of bodies, transforming them into human material or components to be distributed, transferred, or allocated to certain segments of the population. According to these two seemingly disparate examples, it is important to emphasize that certain biotechnologies are used both in animal production and in organ transplants. In principle, biotechnologies are applied to both human and non-human bodies, since they aim to control the biological processes that constitute them. However, the analysis should not only demonstrate that biological processes are effectively controlled through biotechnologies, but also capture and expose the unique transformation of life, in which biotechnologies now play such a significant role. As Foucault pointed out, biopower, while completely invading life, seeks to transform it through power-knowledge relations. However, this analytical approach should also incorporate the notion of ecotechniques, understood as technical management of natural life, since if biopower calculates and manages life, it is crucial to emphasize that the administration and production of life have certain limits if they are not based on technical or technological management of life.[18] Thus, during the twenty-first century, biopower is nothing more than the way in which it completely invades and transforms natural life, but through technical, technological or biotechnological management. Once we have clarified the technical mechanisms employed by biopower, we can examine four case studies to illustrate other ways in which technology connects, reconnects, or restores the connection between humans and nature, whether it does so effectively or simply by promoting such a connection. This is particularly relevant when the imperative of time – which, while justifying the exercise of power, also operates the mechanisms of biopower in other ways – dictates that "life must be saved."

1. In the book chapter titled *How is biodiversity threatened by human activity*, Eric Chivian and Aaron Bernstein state the following: "Humans have already altered to varying degrees nearly half of Earth's land surface, and in the next thirty years, this number will likely rise to about 70 percent."[19] Furthermore, they state:

   > A vast, uncontrolled experiment has been taking place with life on Earth over the past sixty years. Tens of thousands of human-made chemicals have been introduced and distributed around the globe. Many of these clearly have been of enormous benefit to humanity. However, some, including a group of chemicals known collectively as persistent organic pollutants, or POPs, have proven themselves to be significant hazards.[20]

   Regarding these statements, it could be said that the article was originally published in 2008, so, according to its prediction, humans would already be close to altering 70% of the Earth's surface if this trend has continued unabated. However, this modification of the Earth's surface would not be unusual if viewed from a biological perspective, since humans, as a biological species among others, manipulate, construct and transform their environment. But, any organism, as Richard Lewontin and Richard Levins pointed out, that modifies and transforms its environment, can also destroy it.

   > We know rather more, but still far to the little, about how, through their life activities, organisms are the active makers and remakers of their milieu. But a rational political ecological demands that knowledge. One cannot make a sensible environmental politics with the slogan 'Save the Environment,' first, because 'the' environment does not exist and second because every species, not only the human species, is at every moment constructing and destroying the world it inhabits.[21]

If every organism builds its environment, it can also destroy it, then it could be decided that human beings should be an animal among others, an organism among others and a species among others, also construct, configure and destroy their environment. In these biological terms, there is nothing extraneous in knowing that human beings have modified their surrounding environment, but the problem, according to what was written in the second statement, is that the alteration was carried out by this vast and uncontrolled experiment with life on Earth. That is why it cannot simply justify or accept the destruction and devastation that happens in the present. Therefore, it is not a question that human beings are just one biological organism among many, but that their uncontrolled experiments with the Earth are causing destruction. The problem is not that human beings integrate a biological species, but that their science, technology and, ultimately, culture is altering Earth's surface. Hence the issue of the Anthropocene, global warming and pollution that human beings are, with their culture, science and technology, the force that is modifying the planet. Furthermore, we have to assume that this alteration of the Earth's surface, which in turn implies a way of transforming life, is also guided by technical management, understood as the new exercise of biopower related to ecotechnique. Faced with destruction and loss of biodiversity, we must preserve and save life.

2. In 2022, David Bennett, a 50-year-old person who was not suitable for a human transplant, received a genetically modified pig heart. Unfortunately, he passed away a few months later due to a virus that invaded his heart. By means of this paradigmatic example, certain consequences can be noted: (a) contrary to transplants, xenotransplants imply a genetic modification of the animal that will be the organ donor; (b) the limit and availability of organs does not depend on human solidarity and fraternity, because there will be production of organs in farms and laboratories; (c) biotechnology, such as gene editing, can prevent, in advance, the fight against organ rejection by the body, which was one of the main causes of failure in transplants; (d) both the transplant and the xenotransplant demonstrate the productive effect of power charged with saving human life; and (e) the singular transformation of life occurs at the crux between the personal contingency and the technical contingency, which can produce an artificial presence (both that of a human being and that of a genetically modified pig), a science fiction android or a living dead (the first transplant of a genetically modified organ was carried out successfully in the body of a woman with a cerebral death: a living death). Thus, not only culture, science and human technology are modifying the surface of the Earth, but a transformation of life is being carried out through these technical mediations, which seem to produce artificial presences. In this

sense, biotechnology or technique in general artificializes not only human existence but can practically artificialize the Earth's surface.

3. The reintroduction of the Mexican wolf into its environment is another example that demonstrates the way in which biopower, related to ecotechnics, is responsible for invading and transforming life, both while managing it technically. While the Mexican wolf became extinct in wildlife almost 40 years ago, thanks to the binational Mexican wolf conservation and reproduction program, implemented in zoos and breeding centres in Mexico and the United States, it is being saved from extinction. The imperative "life must be saved" is effectively realized here. In this way, the Mexican wolf has escaped extinction. But it is worth highlighting that this was possible because the lives of wolves were technically managed. However, the problem of altering the surface of the Earth and the artificialization of nature emerges, because, in the first instance, the proliferation of wolves in their habitat will produce effects in the trophic chains and in the ecosystem and, in the second instance, technical management implies, as in the case of the genetically modified pig that is transplanted, producing an artificial human entity, the introduction of specimens that have been technically modified. Consequently, the introduction of technically modified wolves would imply the artificialization of the ecosystem. As Eric Katz states:

> Once we inject our intentional designs into a natural system, we no longer have a natural system; we have a garden, a forest plantation, or a farm. We may create a landscape that appears to be a natural wilderness because for various reasons we appreciate the look and experience of wilderness, but this wilderness is a human-made artifact that merely resembles the wilderness produced by natural processes.[22]

In this way, the injection of technically modified wolves would make their habitat a great artifact for the human being. In addition, due to the "injection" of specimens, the singular transformation of life occurs. Then, technique is returning to connect a species, which seemed extinct, with nature.

4. In their manifesto titled *A "Neoenvironmental" Manifesto*, Rob Jackson and Peter Haff identify three problematic issues related to the use of biotechnologies, the proliferation and potential pollution generated by nanotechnology and planetary engineering. These different dimensions can become entirely problematic, which is why they offer three suggestions: promoting the precautionary principle, which implies considering all the negative consequences or effects that a technology, or in this case nanotechnology or planetary engineering, might have; given that the planet is heading towards disaster, it is necessary to catalog and store life on Earth in order to somehow

save it; and finally, in terms of science fiction, reaching a planet where not only could human beings as a species be saved, but where, after colonizing it, the life forms of certain species could be reproduced. These Neoenvironmental suggestions demonstrate a shift in the mechanisms of ecotechnical biopower, which seeks not only to invade and transform, but also to save life by protecting it (the precautionary principle), cataloging and storing it.

The four examples presented illustrate different ways in which human beings connect to, reconnect with, or disconnect from nature: whether by conducting harmful and uncontrollable experiments on nature that alter the Earth's surface (disconnection), by hybridizing genetically modified organisms with humans (connection), by reintroducing Mexican wolves, thought to be extinct, into their natural habitat (reconnection), or by protecting life by cataloging and storing genetic material (connection), even if the ultimate goal is to reproduce it on another planet (a radical, fundamental disconnection).

4. Towards a Symbiopoetic Reconnection with Nature

After reviewing these three aspects in which human beings have historical disconnected, hyperconnected and reconnected with nature through technological means, in this final part the idea is to imagine a form in which, in biopoetic terms, processes of symbiotic life become a historical movement, and, in symbiopoetic terms, human beings understood as holo- and technobionts could restore the connection with other organisms, environments and planet Earth. This proposal seeks to imagine a future in which life still has the capacity and the potential of BIO.

The proposal considers not only the exercise of power *over* life, but the power *of* life. In this case, it is not about continuing to imitate and repeat the behaviours derived from biopolitical practices focused on controlling, instrumentalizing and dominating life, but rather about imitating a certain capacity inscribed in the organisms that have inhabited the Earth for a long time, even before human beings. Imitating these capacities both non-human and pre-human, could open up a more-than-human and post-human future. Likewise, imitating the power of life will lead to transforming life itself into another way, to avoiding extinction by other means, and to cohabiting the planet in symbiotic terms that promote other types of associations and collaborations between organisms that share the benefits. Within this perspective human beings could change their connections with technology and nature, as well as their subjectivity in order to transform the world. Thus, instead of projecting a radical disconnection from the Earth, abandoning the planet, it would be more pertinent to imagine a symbiopoetic reconnection.

If the only way to avert the impending environmental catastrophe is to flee the planet (a radical disconnection), then there is little left to do but resign ourselves and wait for the end. However, rather than passively accepting this adversity, we should explore and re-establish other forms of connection, based on a different exercise of power. This power would no longer be exerted over life but rather would harness the potential of life itself to reproduce and sustain it through alternative means. In other words, it is not simply a matter of ignoring or denying the exercise of biopower, but rather of exercising it in different ways and within different existential dimensions.

In this way, to move beyond the language of violence and subjugation, it is possible to consider a form of imitation or mimicry of the power or potential of life itself. For example, one could consider mimicking the regenerative capacity of axolotls, which stems from the power of their stem cells (which exhibit a double exceptionality: an exception in their origin and an exception in their function), or the symbiotic capacity of organisms. This would suggest that if human beings currently function more like parasites, benefiting from the exploitation of the Earth and other life forms, they could modify their behaviour to become mutualistic symbiotic entities, that is, entities that seek to benefit both themselves and other living things, including the Earth.

Why consider and propose this mimetic approach to the power of life? Aristotle already stated that human beings are the most mimetic animals, although they are not the only ones. This mimetic capacity, related to the psychological and pedagogical spheres, allows individuals to develop a malleable sense of self through the use of models. According to Nidesh Lawtoo, in the article titled *The Plasticity of Mimesis*, this process is formative throughout an individual's life, from childhood to adulthood, meaning that subjectivity is a continuous process, open to new models for imitation. For example, encountering the specific forms of biological discourse motivates the imitation of other models (regenerative or symbiotic power), not merely for the sake of acquiring knowledge or engaging in reflection, but rather to foster life or different life forms. Ultimately, the mimetic capacity acquires political significance because it shapes subjectivity. Therefore, mimesis is not a solipsistic or individualistic exercise, but rather has a political or communal dimension, within which subjectivities are formed. In this sense, using the mimesis of the symbiotic power of life as an example, communal action would transcend any geographical location, where human beings would function as entities that seek not only their own benefit, but also the benefit of all those with whom they interact (mutualism).

In addition to fostering a mimicry of regeneration or (mutualistic) symbiosis, the exercise of biopower could also function in terms of mimicking the life-giving power of the tree form. Typically, when we consider trees, we focus only on their ecosystem functions: what they do, how they function and what their purpose is. In this sense, according to Max Adams and with his book titled *Trees of Life*, trees play a fundamental role in life, due to the many functions they perform and the environmental services they provide, among which are the following: they influence Earth's climate and biodiversity; absorb carbon dioxide and pollutants; release oxygen; recycle water and atmospheric nitrogen; act as natural air conditioners; provide a habitat for millions of species – plants, insects, birds, fungi, mammals and amphibians; stabilize and enrich the soil and help prevent flooding; and by forming forests, create biomes – a large, living organism: a complex web of interdependent and interconnected biological and behavioural relationships.

Now, through the mimetic power of life, there is the possibility that subjectivity could emerge in the manner of the tree-form. Thus, we would have a subjectivity that drives the Earth's climate, as well as its biodiversity. The idea would be to imitate this tree-form so that human beings can positively influence the Earth's climate, because they have already done so, but in a negative way, which has caused global warming. If trees absorb carbon dioxide, a technical or technological method would have to be found to absorb it, such as biomimetic technologies that use microalgae to fix it. In an sf future, the Technosphere could acquire these functions to absorb and fix carbon dioxide, or, if desired, the organic structure of human bodies could be modified to absorb pollutants without dying in the process. An important function of trees is their water recycling. In recent years of drought, rising temperatures, water crises and the implementation of "Zero water day," water recycling has been one of the major problems. Waste and overconsumption can be regulated in other ways by finding alternative uses for water, as well as other ways to recycle and store it. Thus, rather than obeying government mandates to not waste water, we should capture the power of the tree-form to find other uses, as well as other ways of storage and recycling. If trees act as air conditioning and provide a habitat for other species of animals, plants, insects, fungi, etc., human-trees, rather than simply building zoos, aquariums and farms or excluding other species from their cities or, more intimately, from their homes, could create habitats for other species to thrive. They could also do what is necessary to stabilize and enrich the soil, prevent floods, and, even further, create biomes, that is, a large living organism that is itself a relationship of multiple, interdependent and interconnected species with human beings (connection and reconnection). Therefore, it is not about humans becoming trees, but rather about them absorbing, through imitation, the inherent power of tree life in order to live differently, because trees are not only a form of life, but also life-givers, including for human life. As Adams states:

> In a world of plastics, concrete, steel, creeping deserts and dwindling mineral resources, it is worth reminding ourselves of these ever-giving biological, chemical and engineering marvels that can and will sustain so many of our material and aesthetic needs, if only we allow them space, and time.[23]

The idea is not simply to give trees space and time to meet our material needs, but rather to consider how, through this mimetic power, human beings, by emulating the form of a tree, can also satisfy the material and aesthetic needs of trees, forests and the living beings that surround us. By imitating the form of a tree – which would mean capturing the essence of its natural connection to the environment – humanity could achieve a multispecies, technologically-mediated reconnection with nature.

This chapter has sought to demonstrate the technical ways in which human beings can reconnect with nature. This is done through an analysis of, first, the planetary hyperconnectivity generated by satellite technology; second, the connection between the exercise of power and life in general (human and non-human); third, the ways in which ecotechnology, or technical management linked to biopower, facilitates connection and disconnection; and fourth, the reconnection that can be achieved through the imitation of the power of life itself. Moving from exercising power over life to imitating the power of life allows us to grasp the inherent potential of trees, thus fostering a different kind of subjectivity. Rather than oppressing and subjugating life, this approach promotes, through technology, a variety of interconnected relationships with other species. This is particularly relevant as the tree form itself exemplifies certain vital functions, such as influencing the Earth's climate, promoting biodiversity, absorbing carbon dioxide and pollutants, recycling water, providing habitat for numerous other species and creating biomes (large, interconnected ecosystems). In this way, human beings who can emulate the tree form will be able to establish interdependent and interconnected biological and behavioural relationships with multiple species. Furthermore, by capturing the tree's unique way of connecting with nature, this imitation will enable humans to reconnect with nature through alternative means.

Notes

1 See the website https://stuffin.space/ that shows in real time both satellites and space junk circling Earth.

2 For Peter Haff, the Technosphere, understood as the set of technological objects created by humanity, is "the autonomous global system that drives the Anthropocene. It comprises the world's humans and its technological systems, including transportation,

communication, power transmission and financial networks; governments and their associated bureaucracies; military, educational and scientific establishments; religious institutions and political parties; and artistic, political, environmental, cultural and other social movements" (Peter Haff, "The Technosphere and its relation to the Anthropocene" in *The Anthropocene as a Geological Time Unit: A Guide to the Scientific Evidence and Current Debate*, edited by Jan Zalasiewicz, Colin N. Waters, Mark Williams, Colin Summerhayes (Cambridge, United Kingdom: Cambridge University Press, 2019), 138).

3 Russian physicist known for his sf stories and his work on cybernetics.

4 Anatoly Dneprov's, "Crabs on the Island" in *The Molecular Café. Science-fiction stories,* (Moscow: MIR Publishers, 1968), 34.

5 Ibid., 33.

6 Ibidem.

7 Ibid., 43.

8 Ibidem.

9 Ibid., 57–58.

10 Watch, for example, Steve Cutt's short story titled *Man*, in which recreates not only the historical process of dominion of *Man* over nature, but also imagines a possible future world that is flooded with waste and garbage. In the end, *Man* is the King of the techno-waste-garbage-sphere.

11 Oliver López-Corona and Gustavo Magallanes-Guijón, "It Is Not an Anthropocene; It Is Really the Technocene. Names matter in Decision Making Under Planetary Crisis", *Front. Ecol. Evol.* 8, no. 214 (2020): 3. doi:10.3389/fevo.2020.00214

12 Ibidem.

13 Henry David Thoreau, *Walking* (San Francisco: Harper San Francisco, 1994), 36–37.

14 Peter Sloterdijk, "Nature to-be-made. The Crucial Subject of Modern Times" in *Crítica do Contemporâneo: Política*, edited by Rui Mota Cardoso (Portugal and Porto: Fundação Serralves, 2007), 197.

15 Ibid., 199–200.

16 Peter Haff, "The technosphere and its relation to the anthropocene", in *The Anthropocene as a Geological Time Unit: A Guide to the Scientific Evidence and Current Debate,* edited by Jan Zalasiewicz, Colin N. Waters, Mark Williams, Colin Summerhayes (Cambridge, United Kingdom: Cambridge University Press, 2019), 138.

17 For example, see: Birgit Schenider, "Burning worlds of cartography: a critical approach to climate cosmograms of the Anthropocene". *Geo: Geography and Environment* 3, no. 2 (2016), https://doi.org/10.1002/geo2.27

18 See the explanation given by Jean-Luc Nancy about ecotechnique in: Jean-Luc Nancy, "Note sur le term de 'biopolitique'", in *La création du monde ou la mondialisation* (Paris: Galilée, 2002), 137–143.

19 Eric Chivian and Aaron Bernstein, "How is biodiversity threatened by human activity" in *Sustaining Life: How Human Health Depends on Biodiversity* edited by Eric Chivian and Aaron Bernstein (Oxford: Oxford University Press, 2008), 30.

20 Ibid., 54.

21 Richard Lewontin and Richard Levins, "Organism and environment", *Capitalism Nature Socialism* 8, no. 2 (1997), 98. DOI: 10.1080/10455759709358737

22 Eric Katz, "Preserving the distinction between nature and artifact" in *The ideal of nature: Debates about biotechnology and the environment*, edited by G. E. Kaebnick (Baltimore: The Johns Hopkins University Press, 2011), 73.

23 Max Adams, *Trees of Life* (Princeton and Oxford: Princeton University Press, 2019), 9.

Part V

ORGANIZING-INTO-OTHERS

Chapter 9

ORGANIZING-INTO-OTHERS: BIO, ANARCHY AND THE ETHICS OF COLLECTIVE BECOMING

1. Introduction: Why BIO Matters for Organizational Thinking

> All that you touch
> You change.
> All that you change
> Changes you.
> The only lasting truth
> Is Change.
>
> —Octavia E. Butler,
> *Parable of the Sower*

BIO is a daring proposal. It is based on a transformative idea. It required us, as a society, as a polis, to consider changing the way we see reality and humanity in terms of deep interrelationship with nature and ecosystems. Its ambition, obviously, might become its weakness: BIO can be seen by many as a utopia.

This chapter faces this limitation. The general idea is that, to make BIO an idea worth considering, at least for some people, it is necessary to accept practicalities. For example, the way BIO proposes to reorganize our perspectives regarding relationships among humans, groups and governments. BIO requires rethinking how we organize. Which are the bases for viability or even the hope that the way human society has been organized can be changed to make an idea like BIO viable?

The bad news is that such a discussion regarding organizing is strongly attached to the idea that bureaucratic organizations are the *normal* or even the *natural* way to be organized. Bureaucratic organizations, both private and governmental, are based on the impersonal logic of domination and structuration which in turn seem to be the dominant and perhaps only way to organize the world today. In this sense, there might be people who defend that, in practical terms, bureaucracies are the only viable model for organizing. Communitarian organizations, to mention one alternative, have shown their propensity to be captured by strong persons, authoritarian actors who use the communitarian voice to crush difference, plurality and to force consensus. It seems that there is no alternative to hierarchical order and domination.

However, the study of organizing and organizations shows a very different story. Bureaucratic organizations are full of disorder, games of power and resistance. Organization Theory is a well-established field of study and is very well known to be the source of paradoxical case studies where apparently neutral and professional bureaucracies are also composed of communitarian values and even anarchical logics, as Gibson Burrel and Gareth Morgan proposed in *Sociological Paradigms and Organisational Analysis*. Communitarian and anarchic ways of organizing are not just alternative ways of organizing (different from bureaucratic organizations). The fact is more interesting: there is no bureaucracy in practice where anarchic and communitarian logics of organizing do not appear. It is hard to find bureaucracies which can be sustained just by the ideal type of impersonal domination as Paul S. Adler proposes in *The Oxford Handbook of Sociology and Organization Studies*. Emotions, values and performance are substantial ingredients of any formal bureaucratic organization. Shared values, community sentiments appear, sometimes even paradoxically in any bureaucracy. Anarchic ways of organizing are also systematically found in the studies of different organizations (private, governmental or social).

It is the aim of this chapter, then, to show that BIO demands for a reform or even a revolution in the way societies organize, not into a utopia, but something very doable. Doable, given the capacity to crack the illusion of pure hierarchical modern organization (impersonal, technical, neutral in values). Cracking this illusion, there is a strong possibility to find that organizing in the contemporary world implies the need to resolve the mixture of bureaucratic effectiveness with the necessary role of communitarian values and ways of structuring or organizing using and requiring a flexible anarchic way of organization.

In effect, organizations are often imagined as rigid structures: hierarchical, rule-bound and resistant to change. They are designed to control, to manage, to stabilize. Yet the world they inhabit, and the lives they shape, are anything but stable. The specialized literature on organizations has made this very clear at least in the last 70 years: organizations of all types are not based just on order and control. These are elements of any organization. However, this vision is a myth, a very powerful myth: empirical studies show repeatedly that contemporary organizations need to create social and political conditions, both inside and outside, to survive, adapt and even to achieve something. Organizations are more social creatures than mere instruments of domination as Patricia Bromley and John W. Meyer have shown in *Hyper-organization: Global Organizational Expansion.* Then, the question arises: Can organizations become something else? Can they become sites of transformation, care, and collective becoming?

This book has explored the framework of BIO, a conceptual and ethical lens that invites us to rethink life, power and relationality. BIO is not a theory of what life is, but a practice of how life unfolds through transformation, extinction and symbiosis. This book then has offered a rich framework of arguments from speculative fiction, feminist science studies, multispecies ethics and biopolitical critique.

BIO then requires a particular organizational orientation. Organizing under different assumptions, like to think with life, not over it. Organizing not through a linear march toward progress, but a relational unfolding. This implies then a specific discussion for organizations and organizing. For instance, organizations have been seen as instruments, as machines, even though they are composed of people and groups. Organizations are social relationships, apparatuses of sense-making and social constructions of meaning. Nevertheless, the dominant discourse of contemporary organizations keeps thinking of them as hierarchical machines. Organization theory has shown, repeatedly, that these are social creatures, social constructs that are much more than hierarchies and impersonal structures of domination. Organizations are a way of organizing, always becoming in the constant struggle of people and groups to give and create sense and meaning as James G. March defends in *The Handbook of Organizations.*

The central question posed in this chapter is this:

How can organizations engage with BIO in meaningful, ethical and transformative ways?

It asks us to reconsider what organizations are, what they do, and what they could become. It challenges the assumption that organizations must be hierarchical, centralized and control oriented. It requires, then, to think of organizations as symbiotic assemblages, as technosymbionts, as sites of response-ability, as have been proposed earlier in the book.

To do so, we must confront tension. Organizations are often seen as conservative structures built to resist change, that are dominated by powerful and entrenched actors. They are designed to reproduce themselves, to maintain order, to minimize risk. They are governed by the imperative of efficiency, accountability and control. In many ways, they are the antithesis of what BIO would require as a way of organizing.

Can organizations become otherwise? Can they become sites of becoming-with rather than managing over? Can they cultivate new values and meanings like care, justice and transformation, rather than merely pursue growth and profit?

This chapter explores these questions. It draws on the insights of past chapters which have traced the contours of transformation, extinction, symbiosis and BIO.

2. Organizations as Living Systems

In the spirit of BIO, let us turn the attention to organizations, not as static entities or mechanical apparatuses, but as dynamic, evolving social ecosystems. This section explores how organizations can be understood as living systems: self-organizing, symbiotic and multifaceted.

a. Self-organization: life without a central command

In biological systems, self-organization refers to the spontaneous emergence of order from local interactions among components, without centralized control. Flocks of birds, ant colonies and microbial communities all exhibit this property: they respond to internal and external stimuli, adapt to changing conditions and develop new patterns of behavior through feedback and interaction.

Formal organizations, too, can self-organize. While traditional management models emphasize top-down control, planning and predictability, many contemporary organizations (especially in times of crisis or innovation) thrive by embracing decentralized decision-making, emergent strategy and collective intelligence. Teams form organically, roles shift fluidly and solutions arise from the bottom up.

This is not chaos. It is a different kind of order. In BIO terms, it is a form of becoming-with: the organization evolves through its relationships, some of them necessarily conflictive and even, in time, bound to fail.

Self-organization challenges the metaphor of the organization as a machine. Machines require external input, fixed parts and centralized programming. Living systems, by contrast, generate their own energy, repair themselves and evolve. One of the most important findings of the discipline known as Organization Theory or Organization Studies is this: organizations are both instruments and social creatures who resist over instrumentalization and, in doing so, are able to produce new solutions and find different ways to act and innovate.

b. Symbiosis: organizations as holobionts

To understand organizations as living systems is close to seeing them as holobionts or composite entities made up of diverse agents, technologies and environments. In biology, a holobiont is an organism plus all its symbiotic partners.

Organizations are no different. They are not singular entities but assemblages of people, machines, data, cultures and ecosystems. They are shaped by their internal diversity and their external environment, full of other organizations. They host symbiotic relationships, some mutualistic, some parasitic, some commensal.

In the most technical sense, organizations are seen as accountable to their stakeholders. Those agents and groups that are critical for organizational achievement. In this sense, it is not strange for organizational literature to understand the importance of organizational symbiotic dependence. In this sense, stakeholders are the door to ask not only how an organization serves its shareholders, but how it relates to its symbionts, as Christopher Stoney in *Stakeholding: Confusion or Utopia? Mapping the conceptual terrain* observes. Does it nourish its employees or exploit them? Does it collaborate with its environment or extract from it? Does it foster mutual benefit or unilateral gain?

In the BIO framework, organizations are not isolated units but nodes in a web of life. Their health depends on the health of their relationships. Their survival depends on their adaptability. Their flourishing depends on their capacity to become-into-others – to evolve through connection, not just domination.

c. Complexity theory and cybernetics: feedback, emergence, adaptation

To deepen this understanding, let us turn to the theory of complexity and cybernetics: fields that study how systems behave, adapt and evolve. These disciplines offer powerful tools for analyzing organizations as living social systems.

Complexity theory argues that systems are nonlinear, sensitive to initial conditions, and prone to emergent behaviour. Small changes can lead to large effects. Patterns arise from interactions, not instructions. Stability is not the absence of change but the capacity to absorb it, as defined by Douglas Kiel in *Managing Chaos and Complexity in Government a New Paradigm for Managing Change, Innovation, and Organizational Renewal.*

Cybernetics, meanwhile, focuses on feedback loops: the mechanisms by which systems regulate themselves. Positive feedback amplifies change;

negative feedback dampens it. In organizations, feedback loops manifest in performance reviews, market responses, team dynamics and cultural norms. They shape behaviour, reinforce values and give information regarding survival of the organization in its environment.

Together, these theories reveal that organizations are not static structures but dynamic processes. They are constantly learning, adapting and reorganizing. They are shaped by flows of information, energy and meaning. They are not machines: they are social ecosystems, composed of people dealing with their environment. This insight has interesting implications. For example, it suggests that leadership is not about control but about facilitation. Strategy is not about prediction but about responsiveness. Innovation is not about invention but about emergence, risk and transformation.

In BIO terms, complexity and cybernetics help us understand how organizations become-into-others. They show us that transformation is not a top-down mandate but a bottom-up unfolding. They invite us to cultivate conditions for growth, not impose blueprints for change.

If organizations are social-living systems, then they are also sites of becoming. They are not fixed identities but evolving assemblages of persons, groups, conflicts and meanings. They are shaped by their history, their environments and their relationships. They are capable of transformation and change.

To become-into-others, organizations must embrace relationality: they exist due to their interdependence with other organizations and agents. They struggle constantly for adaptability: the capacity to change in response to feedback from their environment. In this sense, organizations are required to be responsible.

BIO then is not necessarily a utopian vision. It can be sustained by very well known practical processes of organizing. Organizations that operate as social living systems need to be resilient and to take risks to innovate. Organizations are well known to be conservative: in search of equilibrium and stability. However, this necessity for survival also implies, not without paradoxes, to search constantly for change and transformation.

3. Organized Anarchies and the Limits of Control

In the early 1970s, Michael Cohen, James G. March and Johan P. Olsen introduced a model of organizational behaviour, they called organized anarchies in *Ambiguity and Choice in Organizations*. These organizations are in several senses, modern bureaucracies: formal structures, impersonal logics of obedience and composed of norms and rules, procedures and regulations. Nevertheless, some of these formal organizations, in order to be understood in practice, show that they are also systems characterized by ambiguous goals, fluid participation and problematic preferences. In these settings, decision making does not follow a linear, rational path. Instead, choices emerge from the chaotic interplay of problems, solutions, participants and opportunities, all swirling together in a metaphorical garbage can.

This model challenged the dominant assumptions of organizational theory at the time. Rather than portraying organizations as rational machines or tightly coordinated hierarchies, Cohen, March and Olsen revealed the messy, unpredictable and often serendipitous nature of organizational life. Their insights resonate deeply with the BIO framework, which embraces complexity, relationality and transformation through situated, responsive practices.

a. Organized anarchies: ambiguity, fluidity and unclear preferences

Organized anarchies are defined by three core characteristics:

1. Problematic Preferences: Goals are often ill-defined, conflicting, or evolving. Members may not agree on what the organization is trying to achieve – or even what success looks like.
2. Unclear Technology: The processes by which decisions are made are not fully understood. Participants may rely on routines, improvization, or trial and error rather than formal procedures.
3. Fluid Participation: People flow in and out of decision-making arenas. Attention is fragmented and involvement varies depending on time, context and interest.

In such an organizational dynamic, decisions are not the result of deliberate planning but the outcome of temporal convergence: when a problem, a solution, a participant and a choice opportunity happen to align. This model does not celebrate chaos for its own sake, but it does acknowledge that ambiguity is endemic, a critical factor which allows negotiation, bargaining, and, lastly, agreement in organizational life.

Rather than resisting this ambiguity, organized anarchies work within it. They do not eliminate uncertainty; they navigate it. They do not impose clarity; they adapt to complexity. This is precisely where BIO finds fertile ground.

b. BIO and the ethics of situated response

The BIO framework is not a blueprint for control. However, it is necessary to build a concept of practical and organizational response-ability: the capacity to respond ethically and attentively to others in specific, situated contexts. It recognizes that transformation does not occur through command and control mechanisms but through relational entanglements and adaptive learning.

In organized anarchies, decision making is often messy, nonlinear and contingent. This messiness is not a failure; it is a condition for becoming. It creates space for experimentation, improvisation and surprise. It allows organizations to evolve not by following a fixed path, but by engaging with the world as it unfolds.

BIO then is a congruent argument in organizational terms: ambiguity is not a pathology, but a constant in organizational life. All organizations need to face ambiguity. However, the solution has been, up to now, to define ambiguity as a pathology and not as a perennial characteristic of any organizational effort. Ambiguity is treated, in discourse as a problem, a pathology to extract and reduce. Nevertheless, the neatness of hierarchical organizations is just an illusion. Proposing that the interplay of different agents in an organization produces pathologies that can be "extracted" through instrumental rationality is not a solution: organizations are arenas full of actors playing games rather than monoliths of blind obedience. People play power, discourses and adaptation games. Power is a critical element for these games. Up to now, the power game has been to legitimize the illusion of the linear, up to bottom use of power as an omnipresent instrument to decision making and action. However, all organizations depend on negotiation. Interdependence, conflict management and ambiguity to negotiate and bargain. A new discourse is needed: one that offers a realistic vision of organizations as arenas full of actors negotiating in highly uncertain environments where new ways of interdependence are critical for the survival of the hope of achieving organizational objectives and missions.

c. Ambiguity as fertile ground for becoming

In traditional organizational theory, ambiguity is often treated as a problem to be solved. It is something to be minimized, clarified, or eliminated. But in BIO, ambiguity is a generative condition. It is the soil in which new forms of organizing can take root.

Ambiguity, there is no doubt, produces problems of coordination and communication. Creates conflicts and challenges. However, ambiguity is productive also, indispensable. It allows for plurality: multiple perspectives, values and possibilities. As plurality is an important source of conflict, it also invites dialogue rather than decree. As Melissa Tyler in *Judith Butler and Organization Theory* observes, conflict opens space for learning partnerships, where knowledge is created through interaction rather than imposed.

Ambiguity is the source of problems, clearly. However, it is also the source of a grille of interaction and interdependence of different actors with different values and perspectives. Ambiguity might produce slower decision-making processes, for instance. Yet, stagnation is not the only destiny of ambiguous social interactions. Here there is no intention to romanticize confusion or indecision. Rather, it is to recognize that certainty is not always ethical, and that control is not always effective. In complex systems (whether ecosystems, organizations, or communities) rigid control can stifle adaptation, suppress diversity and inhibit growth.

Organized anarchies, by loosening control, create room for emergence. Organization then allows organizing in a messy but rich way. Organizations exist and become through interaction, conflict and change, as Haridimos Tsoukas and Robert Chia have studied in *On Organizational Becoming: Rethinking Organizational Change.*

d. Becoming-with in organizational life

To apply BIO in organizational contexts is to shift our focus from control to interdependence. It is to recognize, as Linda Tallberg and Lindsay Hamilton propose in *The Oxford Handbook of Animal Organization Studies*, organizations are composed of diverse agents (human and nonhuman, biological and technological) all entangled in webs of relations.

It is to ask: How do we respond to each other? How do we learn together? How do we shape change in ways that honor difference, nurture mutualism and cultivate deliberation among different?

Organized anarchies offer a model for this kind of inquiry. They show us that decision making does not require perfect clarity, but engagement. In the BIO framework, ambiguity is not the enemy. It is the condition imposed by the diversity of beings involved in a social construct like any organization is.

4. Anarchy as Ethical Practice

In dominant cultural imaginaries, anarchy is often misrepresented as chaos, a breakdown of order, a descent into violence, a rejection of structure. But this caricature obscures the rich ethical and political tradition of anarchism, which offers not disorder, but a different kind of order: one based on horizontal organization and association of diverse rather than similar or in search of unanimity. In the context of BIO, anarchy emerges not as a threat to stability, but as a grammar for an interdependent way of living. A way of organizing that honors diversity, with all its problems but also with all its advantages over a unilateral use of power and domination.

a. Reframing anarchy: horizontal organization and mutual aid

Anarchy, in its philosophical sense, is not the absence of organization but the refusal of imposed hierarchy. It is the belief that people (and by extension, several types of beings) can organize themselves without coercion or centralized control. It is a politics where different actors with a myriad of values can cooperate, experiment and develop, step by step, relationships built on (plural, never unilateral) trust, reciprocity and shared responsibility.

In *Anarchism*, Peter Kropotkin classically argued that mutual aid (not competition) is the driving force of social interaction. This insight resonates deeply with the BIO framework, which emphasizes symbiosis, interdependence and co-creation.

In organizational life, anarchic principles manifest in several phenomena which exist in almost every organization, at least partially: horizontal structures, distributed leadership and consensus-and-conflictive-based decision making. These practices challenge the top-down logic of traditional organizations and invite people to observe that participatory, responsive, and ethical mode of governance is more common than most people think. Real organizational governance is full of possibilities for disagreement, conflict, but also deliberation, discussion and respect for the difference and variability. Real organizations, of all types, have experienced these logics and mechanisms of deliberation and adaptation, in different degrees, around the globe. The issue now is to make such a possibility the hegemonic view, substituting the false idea that vertical domination through bureaucracies is the only practical way to organize.

b. Sympoiesis: beyond commanding-over

Donna Haraway's concept of sympoiesis (making-with) offers a good complement to anarchic thought. Sympoiesis questions the idea of autonomous, self-contained systems and instead embraces distributed agency and relational emergence. It is a way of thinking and being that places response-ability over control.

In sympoiesis, central authority is just one of the possible ways of organizing. The way reality moves is often an ongoing dance of interaction and interdependence. This way of organizing mirrors the anarchic dynamic: a world where beings organize themselves through constant processes of interaction, not always ordered through decree or a single mandate. Coercion then is just one way for producing order. And coercion produces several problems and non-intended consequences as organization theory has shown consistently (for example, Paul Adler in *The Oxford Handbook of Sociology and*

Organization Studies). As Roslyn Wallach Bologh in *Love or greatness: Max Weber and masculine thinking, A feminist inquiry* observes, coercion is just one mechanism of organizations: care has been an important piece of organizing too. A piece generally overseen.

Organizations of all types (bureaucratic, collective, family created, governmental, private, international, non-profit) are dependent on several levels of sympoiesis. No organization can survive without considering the feedback from its environment. Every organization, even the most bureaucratic, can exist in time only through managing diversity stemming from the people that compose every organization. Organizational making sense is a continuing process of bargaining and playing diverse power games, as Michel Crozier and Erhard Friedberg in *Actors and Systems The Politics of Collective Action*, have famously shown.

Sympoiesis and anarchy converge in their commitment to relational ethics. The ambition of BIO perhaps is not to ask how do we live together without domination? But, in other sense, how domination is instrumental to BIO by asking How do we build organizations, capable of achieving goals in an orderly fashion, though also honoring difference, autonomy and interdependence? How do we use domination to make-with, rather than command-over?

c. Anarchic principles and the BIO framework

The BIO framework is, at its heart, an ethical provocation. It calls to reshape kin and world through making adaptability and care fundamental variables for understanding the way we organize government, firms and community organizations.

Anarchic principles align deeply with this call. They might offer a political grammar for BIO.

In practice, this might look like:

- Decentralized governance: where decisions are produced in the most viable transparent way. Sometimes hierarchical, sometimes in a more pluralistic way. Where the process of decision making, its legitimacy, is as important as its effectiveness to be applied.
- Deliberative-based processes: where participation can be open to more actors, looking for ways to make these processes viable and even effective.
- Care infrastructures: where emotional labor, accessibility and well-being are integral part of the main mandates of an organization.
- Conflict transformation: where disagreement is not suppressed, but engaged with a practical sense and even curiosity in order to learn from it.

These practices do not eliminate conflict or complexity. But they offer ways to navigate them without violence or at least searching for the less feasible and justifiable use of domination practices. Domination, the need to produce some obedience in actors to produce cooperation is important, but only one of the ways to create such cooperation.

Anarchy, in this sense, is not a utopia. It is a practice: messy, iterative and grounded in the realities of relational life. It is a way of organizing that honors the multiplicity of needs, desires and perspectives. It is a refusal to reduce life to control, and an embrace of life as a game of relationships and interdependences.

d. Becoming-with

To BIO is in several senses to become-with. To become-with is to become-free, not in the sense of isolation or autonomy, but in the sense of freedom in the middle of interdependence relationships. Anarchy implies understanding that interdependence is always uneven: some have more resources or capabilities than others in different circumstances. Making such a reality transparent is one of the bases to develop anarchic mechanisms: how can people and groups discuss and agree taking into consideration their differences. How can this power be used to produce an even ground for discussion and deliberation? Jürgen Habermas' dream in *The Theory of Communicative Action* requires, perhaps, a solid anarchic theory of organizing.

As Michael Hardt and Antonio Negri explain;

> The most powerful social movements today treat leadership as a dirty word- and for many good reasons. For more than a half of a century activists have rightly criticized how centralized, vertical forms of organization… become fetters to the development of democracy and the full participation of all in political life. And yet, on the other, it is a terrible mistake to translate valid critiques of leadership into a refusal of sustained political organization. to banish verticality only to make a fetish of horizontality and ignore the need of durable social structures.[1]

In this way, anarchic practice becomes a form of symbiopoetics: a way of composing life together, dependent at its minimum. If at all, to hierarchical dynamics.

As we imagine futures where domination becomes a controlled instrument, not an end or a horrible destiny of our social way of organizing, we can also think of futures of multispecies flourishing, ecological justice and organizational transformation. Anarchy offers not just a critique, but a path.

Anarchic way of organizing is a reality. However, it has become a secondary way of organizing. It has been subsumed to the practical and hegemonic bureaucratic way of organization. Developing the theory and practice of anarchic organizations or organized anarchies, becomes an interesting venue to build BIO into a practical and realistic proposition.

5. Case Study: Burning Man as a BIO Organization?

Burning Man is often described as a festival, a gathering, or a cultural event. But to understand its deeper significance, we must see it as something more: an organization. It is a temporary city, a social experiment and a transformative space where self-expression, mutual aid and symbiotic relationships converge. Burning Man is not just a place people go, it is an organized dynamic where people can build, propose and participate in an organized anarchy full of possibilities.

a. Self-expression and emergent order

Burning Man's ethos is grounded in self-expression. Participants are encouraged to bring their full selves (creatively, emotionally, materially) and to take responsibility for their survival in the harsh desert environment of Black Rock City. There are no vendors, no spectators and no passive consumption. Everything is gifted and shared.

This is an extreme example of an organizational anarchy: a refusal, whenever is possible, of imposed hierarchy, an organization based on trust in voluntary association and self-control, and a commitment to horizontal organization whenever it is possible. Neil Shister in *Radical Ritual: How Burning Man changed the World* explains how the festival functions as a temporary autonomous zone where participants build a city from scratch, live together for a week and then dismantle it completely, leaving no trace. It, nevertheless, requires rules, communication, some hierarchical relationships and resources to be managed efficiently.

This organization manages diverse challenges and situations. Its structure allows, nevertheless, to face emergent issues thanks to the limitations of centralized control they have imposed on themselves. The Ten Principles associated with the organization of the festival (including radical inclusion, communal effort, civic responsibility and immediacy) serve as a cultural infrastructure that guides behavior and enables coordination. As Katherine K. Chen documents in her ethnographic study *Enabling Creative Chaos, The Organization Behind the Burning Man Event*, Burning Man's organizers have learned to balance *creative chaos* with enough structure to sustain a temporary city of over 50,000 people.

Chen describes how Burning Man evolved from a small gathering into an unconventional corporation with a multimillion-dollar budget and thousands of volunteers. Yet despite its scale, it retains a commitment to a flexible organization: a system where action is valued over permission and roles are created through initiative rather than assignment.

b. Symbiotic relationships: humans, art, environment and technology

Burning Man is now a formal organization which requires rules and some centralized decision process. However, at the same time. It has been able to maintain its decentralized ethos. It brings together humans, art, environment and technology in ways that blur boundaries between bureaucratic organizations and an organized anarchy. Massive art installations rise from the desert floor, powered by solar panels, built from recycled materials and designed to be touched, climbed and interacted with. Participants collaborate across disciplines (engineering, sculpture, performance, coding) to create experiences that are ephemeral yet quite real.

The principle of *Leave No Trace* ensures that participants minimize their ecological footprint, while the city's circular layout and decentralized infrastructure reflect ecological thinking. Technology, too, plays a vital role. From GPS mapping to solar grids, from digital art to online coordination, Burning Man integrates tools without allowing them to dominate.

All these characteristics does not mean the organization of the Burning Man is free of conflicts and contradictions. As Chen shows, several people have abandoned the project for it becoming too bureaucratic. Others complain that this organized anarchy is too ordered and in different ways, they feel that the organization violates the principle of freedom of domination. Additionally, the organization is well within its environment: not all local authorities and businesses see the Burning Man event as something positive. The challenges of organizing through liberty and plurality are hard and existent. It is important to keep building the practice and theory of anarchic principles of organization where disorder, hierarchies, rules and freedom can be combined in different and novel ways.

Nevertheless, the Burning Man is, in Chen's words, a creative chaos. Burning Man is an organization often messy, unpredictable and even uncomfortable. But it has also shown that it is viable, effective and resilient. It shows that order can emerge from trust, that leadership can be shared and that organizations can exist controlling hierarchy and domination and not becoming captured by such mechanisms.

c. A prototype for BIO organizing

In a world dominated by bureaucracies, algorithms and extractive economies, Burning Man offers a glimpse of an alternative. It shows that people can come together voluntarily, create beauty, care for the planet and leave no trace. It shows that organizations can be relational and transformative.

For those seeking to build BIO-inspired organisations (whether in business, education, activism, or governance) Burning Man provides lessons in:

- Decentralization: Trusting participants to lead, act and innovate.
- Mutualism: Designing systems that benefit several stakeholders.
- Responsiveness: Adapting to changing conditions with creativity and care.
- Embodiment: Honouring the physical, emotional, technological and ecological dimensions of organizing.
- Imagination: Creating spaces where new futures can be rehearsed and realized.

These ideas are part of a small but growing body of scholarship which has begun to challenge the modern view that treats nature as an external domain and human beings as autonomous masters standing apart from it. What is emerging instead is a more relational understanding in which humans are seen as embedded within complex planetary, ecological and multispecies systems.

In *Children of a Modest Star*, Blake and Gilman argue that the "Planetary" names precisely this condition of embeddedness: human life does not unfold outside the Earth system, but within it, entangled with biological, geological and atmospheric processes that disregard political borders[2]. From this perspective, the old conceptual division between humanity and nature appears increasingly untenable, and so too do the institutions built upon it. If our self-understanding is changing, then the ways we organize collective life must also change. The debate is no longer only about improving existing institutions, but about rethinking the very ontological assumptions on which they rest.

This intellectual shift has major implications for how we imagine organization and governance. For instance, in the international arena, international organizations are critical pieces for change and organizing at global level. However, as Laura Zamudio-González and David Arellano-Gault explain in *International Organizational Anarchy*, these important organizations are still seen as pieces dependent on the specific agendas or nations-states – that is, institutions designed to mediate among sovereign states and with a State-centric vision. Perhaps, a good step to discuss is what

Black and Gilman call instead *planetary organizations*: institutions oriented not simply to interstate coordination, but to the governance of problems whose scale and consequences are planetary. Blake and Gilman link this move to a renewed principle of planetary subsidiarity, according to which authority should be allocated to the smallest scale capable of addressing a problem effectively, while recognizing that for certain issues the relevant minimum scale is the planet itself. Their proposal is therefore not a call for a centralized world state, but for an organizational and governance architecture in which local, national, and planetary institutions are reframed according to function and habitability. For a project like *BIO*, this is especially suggestive, because it opens space to argue that new ways of seeing life, nature, and human belonging require not only new ethics, but also new organizational forms.

6. Organizational Dilemmas in the Age of BIO

The BIO framework invites us to reimagine organizational life not as a quest for dominance, but as a relational and ethical practice. In this section, we explore the tensions that define organizational existence in the age of BIO, and the transformative possibilities that emerge when organizations learn to live with, rather than against, these tensions.

a. Navigating core tensions: control vs. emergence

Organizational models are generally built through a dominant image or discourse where all organizations exist due to control, planning, prediction and centralized authority. However, several organizational studies have shown that even the most centralized organization is a living system where emergence and change are perennial. New patterns, ideas and relationships arise not just from command but also from interaction and interdependence. Control seeks stability, something that is crucial for any organization. Nevertheless, emergence appears and organizations can change and adapt to new circumstances. Sometimes, organizations exploit what has been working for them; however, very often it is time to explore new ways and experiments, as March in *Explorations in Organizations* shows. Exploration and innovation are not free of risk, though. How can an organization then change? Through their people, groups and coalitions. An organization then is not a monolith as the dominant discourse about management generally defends. Organizations are better imagined with another metaphor: the arena. As David Arellano-Gault and Arturo Del Castillo propose in *Promises and Perils of Compliance*, arenas are the space where interactions, games and negotiations happen.

Real organizations are always confronting emergent situations due to their dynamic life.

In the BIO framework, emergence is not a threat but an intrinsic characteristic of organizational action. In this sense, the capability to change and adapt to new circumstances and values is the space where organizing happens. Organizations evolve through surprise and relational dynamics. To embrace emergence, organizations can loosen their grip on certainty and cultivate conditions for creativity, responsiveness and growth. Even if doing so implies a strong modification of values and old certainties.

b. Navigating core tensions: efficiency vs. care

Efficiency is often framed as the highest organizational value: do more with less, faster, cheaper, better. Pursuing efficiency as the only and most important value, however, creates its own problems. It can flatten complexity but also erase nuance. Most importantly, very often efficiency has been a way of justifying exploitation and sacrifice of the well-being of people and the environment. There is another value that has appeared slowly but steadily, as another critical organizational value: care. Care by contrast, is slow, situated and relational. It asks: What do people need? What do ecosystems require?

In the age of BIO, care is becoming a necessity. Organizations can learn to balance efficiency with attentiveness, productivity with attention. The current situation perhaps needs to recognize that care is not antithetical to performance, but foundational to sustainability. Without care, efficiency becomes extraction. With care, efficiency requires to invent new forms of governance where different actors can offer input. Considering care as a normal element of management, efficiency can develop ways of governance able to learn from the decentralized intelligence that exists in the organization and outside it.

c. Navigating core tensions: hierarchy vs. kinship

Hierarchy organizes power through a vertical framework. Kinship aspires to organize power through the most feasible participation possible. In hierarchical systems, relationships are instrumental. In kinship systems, relationships are spaces where people interact and produce the bases for cooperation and sense giving, as proposed by Karl Weick in *Making Sense of the Organization*.

BIO calls for a reconfiguration of organizational relationships – from domination to communication, from command to agreement. Kinship does not mean the absence of structure, but the presence of mutual respect and distributed agency. It means recognizing that every member of an organization (human or non-human, technical or biological) is part of a web of interdependence.

d. Navigating core tensions: metrics vs. meaning

Metrics quantify performance. They offer clarity, comparability and accountability. But they can also obscure other important matters: purpose, connection, persons' needs and interests. Sense making in an organization is a human, group and collective process. Sense making is produced through stories, relationships and lived experience.

Organizations need learning to hold both: honour meaning as the heart of their work even for the way the organization develops metrics. In BIO terms, meaning emerges from becoming, from the ways organizations respond to change and cultivate care.

e. Practices for transformation

Symbiotic design reframes organizations as holobionts: assemblages of diverse agents, technologies and environments. In this sense, the organizational question of how to survive perhaps is changing in substantial ways. The question now becomes more related to survival not as an isolated piece, as disposable instrument managed by few people. Now the question might be how do we support mutual flourishing since the organization is a particular holobiont? A holobiont responsible for the well-being of its parts, including the environment.

This means designing for interdependence. It means creating feedback loops. It means honoring the needs of all symbionts (employees, communities, ecosystems, technologies) and recognizing that organizational health depends on relational health.

Power is often understood as control: the ability to direct, dominate, or decide. But in BIO, power is relational. It flows through connection, not command. It is not a possession, but a process – a capacity to create, to sustain.

Organizations can reconfigure power by accepting that the intelligence in an organization is distributed through all the different groups and parts that compose it. The change of paradigm is possible: flattening hierarchies, fostering collaboration and creating spaces for emergence are real processes that several organizations have experimented through time. In other words, these strategies of shared power are more common than some people might think: the metaphor of mechanical organizations has been shown wrong since the early times of organization studies. Why then, the main image and discourse of organizations are based in hierarchical control and the idea that it is a monolith and not a living arena of games and interdependences? There is no doubt that there is a need to develop a more detailed and empirical theory of organized anarchies, as has been discussed earlier. It is time to overthrow the dominant metaphor of bureaucracies as the only organizational alternative at hand. Organized anarchies are a very realistic and existent way of organizing, one that enables the creation of new ideas and possibilities for finding ways of organizing ourselves.

f. BIO and the challenges of governance and public policy

Up to now, it seems clear that the discussion developed here can be expanded by placing BIO in conversation with other philosophical and political traditions that illuminate its organizational implications more explicitly. One particularly fertile dialogue is with Joan Tronto's ethic of care. Tronto's importance for the present argument lies in her refusal to reduce care to a private virtue, an emotional disposition, or a secondary moral supplement to institutional life. Particularly, what Tronto defines as *caring with*, is a political and practical activity: a way of maintaining and repairing the world so that life can be sustained as well as possible. Caring with implies plurality, communication, respect, trust and solidarity. Read in this light, BIO is not only a speculative or ethical invitation to think relationally about life, extinction, symbiosis, and becoming. It is also a way of rethinking organizing itself as the collective labor of sustaining interdependent worlds as Tronto discusses in *Caring democracy: Markets, equality, and justice.*

Tronto's contribution can be pushed even further. Her account of democratic care insists that societies are judged not only by abstract commitments to justice, but by how they distribute caring responsibilities, how they recognize dependency and vulnerability, and how they make those responsibilities visible in institutional life. Care, in this sense, is not benevolence. It is a political criterion for evaluating whether institutions are organized to perceive needs, assume responsibility, deliver competent responses, and remain attentive to the voices of those who receive and provide care. Read alongside BIO, this insight is especially powerful because it prevents relationality from becoming a merely poetic language. It anchors becoming-with in concrete questions of labour, responsibility, asymmetry, and institutional design.

At the same time, Tronto's framework also reveals one of the limitations that BIO and related relational approaches must confront. The ethic of care is normatively compelling, but it does not by itself tell us enough about how organizations should be structured so that care can be stabilized under conditions of scale, conflict, scarcity, and unequal power. To say that institutions should care is not yet to explain how authority should be distributed, how accountability should be organized, how trade-offs should be decided, or how care should be protected from bureaucratic capture and symbolic appropriation. For BIO to become a viable organizational and governmental orientation, the discussion must therefore move beyond moral critique toward more operational questions of governance: coordination mechanisms, jurisdictional design, conflict resolution, budgeting, and the institutional conditions under which care can become durable rather than aspirational.

In any case, if organizations are not machines but living social systems, if they are arenas of plurality, ambiguity, negotiation, and responsibility, then care cannot be treated as external to organizational life. It is one of the constitutive practices through which organizations endure, adapt, and preserve the conditions of their own legitimacy. Care implies a radical move from organizing as understood as extraction or reduction of life to control.

A second useful dialogue is with pragmatism, especially with John Dewey's reflections on public and collective inquiry. Dewey argued in *The Public and Its Problems* that publics emerge when the indirect consequences of associated action become sufficiently extensive and serious that they require reflection, institutional mediation, and collective response. This idea is valuable here because it links organizational life to broader societal frameworks without treating society as a separate or higher-order level detached from organizations. Publics are formed around the consequences of organized action. Problems become public when patterns of interdependence generate effects that exceed private control and demand common attention. In that sense, governance and policy are not separate from organizational analysis. They are modes through which societies organize inquiry, responsibility, and intervention around shared problems.

This point becomes even clearer if we consider public governance not as a sphere external to organizations, but as one of the principal ways in which collective interdependence is organized. Governance is not only what governments do. It is the provisional coordination of heterogeneous actors, institutions, knowledges, infrastructures and environments under conditions of plurality and uncertainty. Seen from a BIO perspective, governance is an organizational accomplishment: it depends on relations among agencies, communities, firms, experts, technologies, and ecological systems. Its success therefore cannot be judged solely by command capacity, procedural compliance, or speed of decision. It must also be judged by its ability to perceive needs, process difference, remain open to feedback, and revise its courses of action in light of consequences. In Tronto's terms, the quality of governance is inseparable from the quality of care institutionalized within it.

Something similar can be said about public policy. Policy should not be imagined as a purely technical decision that is made in abstraction and later implemented by organizations. Policy is constituted organizationally from the outset. It emerges through practices of classification, deliberation, budgeting, negotiation, interpretation, monitoring, and revision. Policies are shaped not only by state elites or formal decision makers, but also by street-level actors, professionals, communities, digital infrastructures, and the wider environments affected by them. A BIO approach makes visible that policy is always relational: it organizes who is heard, what counts as relevant knowledge, which lives are protected, which sacrifices are normalized, and which futures are rendered imaginable.

From here, Elinor Ostrom's work becomes especially suggestive. In *Governing the commons: The evolution of institutions for collective action* Ostrom demonstrated that collective action cannot be reduced to the dualism of centralized state control versus atomized market exchange. Between those poles there exists a wide range of institutional arrangements based on nested, overlapping, and partially autonomous centers of decision. This polycentric sensibility is important for BIO because it shows, in organizational and institutional terms, that coordination need not depend on a single hierarchical center. Complex collective problems often require forms of organization capable of distributed learning, mutual adjustment, and rule revision across multiple scales. This is not the disappearance of authority, but its reconfiguration within systems that acknowledge interdependence and the limits of unilateral control.

A broader relational horizon can be drawn from Arturo Escobar's pluriversal perspective. Escobar proposes in Designs for the Pluriverse: Radical Interdependence, Autonomy, and the making of worlds that to organize is always, in some sense, to compose a world. Organizational forms, governance arrangements, and policy instruments are never neutral containers. They privilege specific relations, temporalities, and ways of inhabiting the earth. BIO gains depth when interpreted through this lens, because it becomes possible to see that the central question is not only how organizations function, but what kinds of worlds they help enact. This is precisely why governance and policy should not be added to the discussion as external implications. They are among the main arenas in which organizational world-making takes place.

Escobar's importance lies also in his insistence that design is ontological: every institutional arrangement helps bring particular realities into being while marginalizing others. The pluriverse is therefore not a celebration of fragmentation for its own sake, but an argument that many worlds already exist and struggle to persist within asymmetric fields of power. From this standpoint, governance is never simply administrative. It is always a dispute over which relations are protected, which knowledges are rendered authoritative, which temporalities are privileged, and which ways of dwelling together are made institutionally possible. BIO converges with this perspective insofar as both reject the fiction of neutral organization and foreground the world-making consequences of organized action.

Yet Escobar's *pluriversal* horizon also encounters an important practical limit when translated into organizational analysis. Its critical force is strongest at the level of ontology, design and political imagination, but less developed when the question becomes how heterogeneous worlds are to be coordinated in everyday institutions without being reduced to a single logic. Once the discussion enters the terrain of public governance, difficult problems immediately appear: how to design decision rules across ontological difference, how to mediate disputes among actors whose values are not commensurable, how to link local autonomy with large-scale infrastructures, and how to prevent pluriversality from becoming administratively unworkable or politically co-opted. BIO, if it is to be realistic, needs to learn how to translate *pluriversal* sensitivity into institutional arrangements, governance architectures, and organizational practices capable of operating amid conflict, asymmetry and interdependence.

The broader implication is straightforward but important. If BIO asks us to think with life rather than over it, then public governance and public policy need to be conceived as organizational practices of collective care and distributed intelligence. This does not require romanticizing horizontality or denying the need for coordination, hierarchy, or authority. It requires something more demanding: institutions capable of using authority without confusing it with blind domination, of organizing action without silencing plurality, and of pursuing effectiveness without severing it from care. Under these conditions, governance, policy, and organizing appear not as separate domains, but as intertwined practices through which societies attempt to sustain life in common.

7. Final thoughts: Towards a BIO Organizational Ethos

Organizations, like ecosystems, are always in flux. They are shaped by the relationships they host and the futures they imagine. In the BIO framework, organizations are not machines to be optimized, but are living social systems, capable of transformation, adaptation and even care. They are not merely sites of production or control, but spaces of becoming-with.

To embrace a BIO organizational ethos is to reimagine what organizations are. It is to see them not as mechanical monoliths of hierarchy, but as arenas for the interplay of people, groups, ideas and a plurality of values. A space where people learn to live together and with others (including non-human animals) in mutual responsiveness.

Traditional organizations often operate in a *doing-for* mode: leaders do for employees, companies do for customers, institutions do for society. This model assumes separation, hierarchy and unilateral action. In BIO organizations become holobionts or assemblages of diverse beings, technologies and environments. They learn to listen, to adapt, to respond. They become sites of sympoiesis where value is discussed and created, not extracted.

The main doubt raised by a BIO organizational ethos is clear: plurality generates conflict, and conflict often slows and even stalls decision-making. In today's world of consumption and the accelerated pursuit of profit, these ideas may seem utopian, even impractical. Yet research on organizational fiascos and disasters – whether labelled decision traps as in Arellano-Gault *Las trampas de la decision*, or as *absurd decisions* in Christian Morel in *Les décisions absurdes* – demonstrates something crucial: organizations are frequent sources of irrational and destructive behaviour, leading not only themselves but also their stakeholders, communities and even entire populations into disaster.

These studies consistently point to a central cause: stubborn decision-making that refuses to confront plural voices or incorporate diverse information and critique. Rationalist and technocratic decision-making styles tend to share a peculiar weakness: an active effort to avoid discussion, silence critique and dismiss the multiplicity of voices that shape the organizational arena. When critique and plurality are systematically excluded, failure and fiasco become highly probable.

The lesson is this: critique is uncomfortable and plurality (rather than presumed homogeneity) is indeed problematic and conflictive. Critique can also be manipulated politically to overwhelm adversaries. But despite these costs, critique introduces differences, novelty and corrective information into the system. It is indispensable for redirecting misguided paths and avoiding disaster. Plurality, for its part, acts as a safeguard against hubris and self-deception, and it provides access to distributed intelligence – since knowledge and insight exist in many corners of an organization, not in a single hierarchical center.

BIO organizing does not mean rejecting hierarchy, efficiency, or power in favour of a utopian vision of conflict free, fully horizontal organizations. The aim is different: to foster organizations capable of confronting conflict, recognizing that plurality, doubt, critique and the constant reevaluation of values and decisions constitute a deeply rational and powerful way of organizing. This approach, like any other, comes with costs. But it embraces one of the most enduring lessons of organization theory: there is no single way to organize.

More than ever, it is evident that all organizations are living social systems, full of plural voices and an urgent need to find new ways of coping with uncertainty. Clinging to the belief that there is only one rationality for organizing is, increasingly, a mistake in need of correction. The task is to open the door to discussion, experimentation and the exploration of different forms of organizing. Contemporary literature on organization theory is clear: "…organizations are depicted as coherent, autonomous, and rational actors,"[3] when in reality they are contradictory social creatures in constant search for adaptation, innovation and survival. This is why, as several studies have shown, organizations use a large set of alternatives, from hierarchy to deliberation, from centralization to new ways of horizontal governance. From the illusion of order to a reality of organized anarchy.

Yes, organizational anarchy, as has been developed in this chapter, can be slow, and requires new ways of governance and deliberation. It might challenge several ways to obtain outcomes efficiently. However, the literature on decision traps shows that without taking in consideration the plurality of interests and actors.

BIO would require experimenting new and daring ways to deliberate, discuss and create mutual understanding based on the diversity of ideas and values. Experiments like those of the *Good Judgement Project* by Philip Tetlock and Dan Gardner in *Super forecasting* or DebateCAFE by Ken Goldberg and Mo Zhou in *DebateCAFE v1.0: Incentivizing Articulation and Consideration of Adversarial Arguments*. Both proposals are excellent examples that diversity of ideas and values can be processed viably to be treated respectfully and rationally through deliberation.

This shift also requires humility. It requires organizations of all types to ask: Who are we becoming with? What are we co-building? What kind of world are we helping to create understanding that we all (and future generations) must have a life in such a world?

To conclude this chapter is to extend an invitation. Let us imagine how to organize, not to dominate, but to shape worlds worth living in. Organizations have always been dependent on organizing. Organizing is a fluid process of sense making, deliberation, and playing games in the organizational arena. BIO organizing, then, is not a utopia: its practices have long existed and have been practiced at different levels and intensities through time.

Notes

1 Michael Hardt and Antonio Negri, *Assembly*, (Oxford: Oxford University Press, 2019), XIV.

2 This switch global to planetary perspectives, can also be thought with the proposal of historicist Dipesh Chakrabarty who, in his book *The Climate History in a Planetary Age* argues, that planetary differs from global perspectives in the sense that the formers are not humanocentric, and invite us to think about various life forms and the conditions in which life on the planet can persist. Conditions which are embedded to the histories of humans' systems, organizations and hierarchical constructions.

3 Patricia Bromley and John Meyer, *Hyper-organization*, (Oxford: Oxford University Press, 2015), 157.

Postscript

DISCONNECTION VS. RECONNECTION

Jane Goodall, a renowned primatologist and animal enthusiast, left this world recently but her legacy survives in the myriad projects and institutes she founded and worked throughout her life. She was a pioneering and creative explorer of nature, principally of chimpanzee behaviour, who left us marvellous documentaries that inspired generations towards a reconnection to the natural world. Although she might be a controversial figure in many ways,[1] we want to honour her legacy through this book by remembering her groundbreaking epistemology and methodology in trying to understand animal behaviour and environmental sciences through an empathic and affective approach that broke through many standards of her time.

Before her death Goodall left the generations who inhabit this planet a saying that encompasses, so to speak, her life experience generated by living with other primates. The saying goes as follows: "We have lost our connection with the natural world. And if you don't have a connection with something, you can't care for it, love it, or understand it."[2] We believe that Goodall's legacy, captured in this saying, can help elucidate the central idea of BIO's proposal.

However, it's worth clarifying some ideas that revolve around Goodall's statement and, in any case, what we are committed to in this book and in our proposal. We de facto accept that we have lost connection with the natural world. This is especially true when we live in hyper-technological cities that integrate nature as an ornament or a kind of sustainable cosmetic, whose sole function is to beautify the urban landscape, both in public and private spaces. Nature is an ornament superimposed on the technical. In this sense, first, it would not be a question of determining why we have lost connection, but rather what we have connected to such a degree that it has led us to a disconnection from nature. The answer could be evident in the times in which we inhabit a technosphere, which in some ways overflows the biosphere; second, it would be necessary to insist on creatively generating a reconnection with nature. Although human beings have already disconnected from nature, there is the possibility of establishing a way to reconnect, which will not be visible to the naked eye, but rather through the technical or technological means at hand.

Disconnection can be overcome once human beings are willing to reconnect with nature, which somehow continues to exist, despite everything and in its improbability, beyond the human. The diagnosis of disconnection is meaningless if we don't think creatively about ways to reconnect. The question arises: how to reconnect? Following Goodall, the first way would be through love. But what is love? In these present times, in which a whole range of hatreds have been identified, ranging from xenophobia to transphobia, we can cross the threshold by knowing what hatred is in order to approach love. If hatred is separation, estrangement and ultimately destruction, love would be union, encounter and ultimately a creative process. In this way, reconnection would have to be based on a creative process. If love is what reconnection is about when it comes to reconnection, then it would have to be a creative reconnection with the natural world.

Now, what about care? We normally use the expression "take care" as we say goodbye to people, in this context, care means: good intentions, that this other entity may do well, a kind of concern, or, at best, that this other entity does everything possible to maintain its existence. But care has a deeper dimension, namely, that care, or the way of taking care of oneself, is related to exercises, practices, or training that not only procure or benefit one's own life, but also do everything possible for a life to achieve fulfillment and, in any case, the excellence of being and existing. Consequently, reconnecting with nature in terms of care would mean carrying out a whole series of exercises, practices and training that, accompanied by all the techniques it might entail, allow for or foster the fullness and excellence of living beings. In this case, care goes beyond the forms of domination and subjugation that would lead to fulfillment of the commands of a sovereign will.

Finally, understanding the reason and purpose of reconnection entails understanding the processes by which other living beings are able to become. Furthermore, it involves understanding that living beings do not become what they are on their own (as if they were completely closed systems), but require, in the very intimate sense of coexistence, the existence of other living beings that, through interaction, become what they are or can become something different over time.

In line with these ideas derived from Goodall's statement, BIO is a proposal that seeks to establish a reconnection with other lives centred on a creative process, in which the connected parties can reach their levels of fulfillment and excellence, avoiding any type of domination or submission. That modes of domination and submission will eventually emerge is something that cannot be avoided, but the venture is to rely on this strategy in a minimalistic fashion. The way to free ourselves from these power relations proceeds precisely through an understanding of the processes by which life has taken possession of this planet, to the point of maintaining it for several billion years. In this way, reconnection founded on love, care and an understanding of life can lead to a different perspective on the future, as long as we begin in this present to reconnect in multiple ways with life in general.

After all, future generations, both those beginning to grow up on this somewhat damaged planet and those who arrive without fully understanding the environmental conditions they have fallen into and what they will have to face, will appreciate the efforts their human ancestors made to reconnect with nature in multiple ways. Rather than dealing with environmental catastrophe, these generations can become givers of life… of an improbable life that will foster other connections aimed at ensuring that life continues on this planet.

Let the reconnection process begin.

In loving and grateful memory of the amazing "Trimates"
Dian Fossey (1932–1985)
Jane Goodall (1934–2025)
Birutė Galdikas (1946–2026)

Notes

1 Jane Godall's work has been examined because of her links with fortress conservation practices (the displacement of indigenous people to create wildlife sanctuaries) and in general, her links with colonial practices in conservation like "green colonialism" and insensitivity to racial dynamics. Recently there's also been controversy of her Jane Goodall Institute in Israel, located in Palestinian territory. Even though through her Institute she has reproved racism, acknowledged her privileges throughout her career, and condemned the war as a genocide; her links to these practices, institutions, and governments have made her a controversial figure, which we believe important to highlight.

2 Jane Goodall, "La importancia de proteger a los primates", interview, 30 September, 2025, N+, https://youtu.be/N_1l-R36OE8?si=Z-RUL3UAhbjSpcFW.

REFERENCES

Introduction

Butler, Octavia E. *Octavia E. Butler: The Last Interview and Other Conversations.* Melville: Melville House, 2023.

Calarco, Matthew. *Zoographies: The Question of the Animal from Heidegger to Derrida.* New York: Columbia University Press, 2008.

Cimatti, Felice and Salzani, Carlo (eds.), The Biopolitical Animal. Edinburgh: Edinburgh University Press, 2024.

Gabardi, Wayne. *The Next Social Contract: Animals, the Anthropocene, and Biopolitics.* Philadelphia: Temple University Press, 2017.

Hardt, Michael and Antonio, Negri. *Assembly.* Oxford: Oxford University Press, 2019.

Haraway, Donna. *Staying with the trouble: Making kin in the Chthulucene.* Durham: Duke University Press, 2016.

Jacobs, Michael and Young, Bob. "Dinosaurs." 1991. IMDb. April 26, 1991. https://www.imdb.com/es/title/tt0101081/.

Ko, Aph and Ko, Syl. *Aphro-ism: Essays on Pop Culture, Feminism, and Black Veganism from Two Sisters.* New York: Lantern, 2017.

Le Guin, Ursula K. *The Birthday of the World and Other Stories.* New York: Perennial Harper, 2003.

_____. *The Carrier Bag Theory of Fiction.* Cornwall: TJ Books, 2019.

Malabou, Catherine. *Ontology of the Accident: An Essay on Destructive Plasticity.* Translated by Carolyn Shread. Malden: Polity Press, 2012.

Taylor, Sunaura. *Beasts of Burden: animal and disability liberation.* New York: The New Press, 2016.

Vint, Sherryl. *Animal Alterity: Science Fiction and the Question of the Animal.* Cambridge: Liverpool University Press, 2010.

Wolfe, Cary. *Before the Law: Humans and Other Animals in a Biopolitical Frame.* Chicago: University of Chicago Press, 2012.

Woolf, Virginia. *Three Guineas.* New York: Harvest Books, 2016.

Chapter 1

Atwood, Margaret. *Oryx and Crake.* New York: Anchor Books, 2004.

Bauer, Liza. *Livestock and Literature: Reimagining Postanimal Companion Species.* Sheffield: Palgrave Macmillan, 2024.

Butler, Octavia. Wild Seed. New York: Grand Central, 1980.

_____. Fledging. New York: Grand Central, 2005.
_____. *Parable of the Sower.* London: Headline, 2019.
Haraway, Donna. "Situated Knowledges: The question in feminism and the privilege of partial perspective". *Feminist Studies* 14, no. 3 (1988): 575–599. https://doi.org/10.2307/3178066.
_____. Simians, Cyborgs and Women: The Reinvention of Nature. New York: Taylor and Francis, 2013.
_____. *Staying with the trouble: Making kin in the Chthulucene.* Durham: Duke University Press, 2016.
Ko, Aph. *Racism as Zoological Witchcraft: A guide to getting out.* New York: Lantern Publishing, 2021.
Kozo-Polyansky, Boris Mikhaylovich. *Symbiogenesis: a new principle of evolution.* Edited by Victor Fet and Lynn Margulis. Cambridge: Harvard University Press, 2010.
Le Guin, Ursula K. *Dancing on the Edge of the World: Thoughts on Words, Women, Places.* New York: Grove Press, 1989.
Margulis, Lynn and Sagan, Dorian. *Acquiring Genomes: A Theory on the Origin of Species.* New York: Basic Books, 2002.
Oyěwùmí, Oyèrónkẹ. *The Invention of Women: Making an African sense of Western Gender Discourses.* Minneapolis: University of Minnesota Press, 1997.
Spillers, Hortence. "Mama's Baby, Papa's Maybe: An American Grammar Book". *Diacritics* 17, no. 2 (1987): 64–81. https://doi.org/10.2307/464747.
Taylor, Sunaura. *Beasts of Burden: Animal and Disability Liberation.* New York: The New Press, 2017.
Wynter, Sylvia. "1492: A New World Perspective," in Lawrence Hyat, Vera and Nettleford Rex, Race, discourse and the origin of the Americas. Washington: Smithsonian Institution Press, 1992.

Chapter 2

Agamben, Giorgio. *The Use of Bodies. Homo Sacer IV, 2.* Stanford: Stanford University Press, 2016.
Adams, Douglas. The Hitchhicker's Guide to the Galaxy. London: Gollancz, 2002.
Bloomberg Originals. 2024. "How Tech Is Breaking the Rules of Biology | Posthuman with Emily Chang." YouTube. December 2, 2024. https://www.youtube.com/watch?v=88DPXE1thw4.
Cleland, Carol and Chyba, Christopher. "Does 'life' have a definition?". In *The Nature of Life. Classical and Contemporary Perspectives from Philosophy and Sciences,* edited by Mark A. Bedau and Carol E. Cleland, 326–339. Cambridge: Cambridge University Press, 2010.
Darwin, Charles. *The Origin of Species.* Hertfordshire: Wordsworth Editions Limited, 1998.
Foucault, Michel. *The History of Sexuality: an introduction.* New York: Pantheon Books, 1978.
_____. "The Crisis of Medicine or the Crisis of Antimedicine?". *Foucault Studies* 1, (2004): 5–19. https://doi.org/10.22439/FS.V0I1.562.
_____. *The Order of Things: An Archaeology of the Human Sciences.* London: Routledge, 2005.
Groys, Boris. "Art in the Age of Bio-politics: From Artwork to Art Documentation". In *Art Power,* 53–65. Cambridge: The Mit Press, 2008.

Lamarck, Jean-Baptiste. *Zoological Philosophy. An Exposition with Regard to the Natural History of Animals.* Chicago: The University Chicago Press, 1984.

Machery, Edouard. "Why I stopped worrying about the definition of life... and why you should as well". *Synthesis* 185, (2011): 145–164. https://doi.org/10.1007/s11229-011-9880-1.

Mariscal, Carlos and Doolittle, W. Ford. "Life and life only: a radical alternative to life definitionism". *Synthesis* 197, (2020): 2975–2989. https://doi.org/10.1007/s11229-018-1852-2.

Simmel, Georg. *The View of Life. Four Metaphysical Essays.* Chicago: The Chicago University Press, 2010.

Thacker, Eugene. "Biophilosophy for the 21st Century". In *Resisting Bio-politics. Philosophical, Political, and Performative Strategies*, edited by S. E. Wilmer and Androné Zukauskaité, 123–134. New York: Routledge, 2016.

Van Camp, Nathan. *Redesigning Life. Eugenics, Bio-politics, and the Challenge of the Techno-Human Condition.* Bruxelles: Peter Lang, 2015.

Chapter 3

Biasetti Pierfrancesco, Hildebrandt Thomas B., Göritz Frank, Holtzel Susanne, Stejskal Jan, Galli Cesare, Čižmàr Daniel, Simone Raffaella, Seet Steven, de Mori Barbara. "Ethics at the Edge of Extinction: Assisted Reproductive Technologies (ART) in the Conservation of the Northern White Rhino". *Journal of Agricultural and Environmental Ethics* 38, no. 7, (2025): 1–22. https://doi.org/10.1007/s10806-024-09943-z.

Bradshaw, Karen. *Wildlife as Property Owners: A New Conception of Animal Rights.* Chicago: Chicago University Press, 2020.

Braidotti, Rosi. *The Posthuman.* Cambridge: Polity, 2013.

Butler, Octavia. Dawn. London: Headline Publishing Books, 2022.

d'Eaubonne, Françoise. *Feminism or Death.* New York: Verso, 2022.

Donaldson, Sue and Kymlicka Will. Zoopolis: A Political Theory of Animal Rights. Oxford: Oxford University Press, 2011.

Haraway, Donna. "Situated Knowledges: The Science Question in Feminism and the Privilege of Partial Perspective". *Feminist Studies* 14, no. 3 (1988): 575–599. https://doi.org/10.2307/3178066.

_____. When Species Meet. Minneapolis: University of Minesotta Press, 2008.

_____. *Staying with the trouble: Making kin in the Chthulucene.* Durham: Duke University Press, 2016.

Kirksey, Eben. "Chemosociality in Multispecies Worlds: Endangered Frogs and Toxic Possibilities in Sydney". *Environmental Humanities* 12, no. 1 (2020): 23–39. https://doi.org/10.1215/22011919-8142198.

Noske, Barbara. *Human and Other Animals.* London: Pluto Press, 1989.

Noske, Barbara. "Two Movements and Human-Animal Continuity: Positions, Assumptions, Contradictions". *Animal Liberation Philosophy and Policy Journal* 2, no.1 (2004): 15–26.

Spatola, Gabriella J., Buckley Reuben M., Dillon Megan, Dutrow Emily V., Betz Jennifer A., Pilot Małgorzata, Parker Heidi G., Bogdanowicz Wiesław, Thomas Rachel, Chyzhevskyi Ihor, Milinevsky Gennadi, Kleiman Norman, Breen Matthew, Ostrander Elaine A., and Mousseau Timothy, "The dogs of Chernobyl: Demographic insights into populations inhabiting the nuclear exclusion zone". *Science Advances* 9, (2023): 1–16. https://doi.org/10.1126/sciadv.ade2537.

Taylor, Sunaura. *Disabled Ecologies: Lessons from a Wounded Desert.* California: University of California Press, 2024.

Turnbull, Jonathon. "Checkpoint dogs Photovoicing canine companionship in the Chernobyl Exclusion Zone". *Anthropology Today* 36, no. 6, (2020): 21–24. https://doi.org/10.1111/1467-8322.12620.

Unferth, Deb Olin. *Barn 8.* Minnesota: Graywolf Press, 2020.

Chapter 4

Ash, Jenny, dir. Extinct: Ep. 6, The Tasmanian Tiger. 2001. IMDb. October 30, 2001. https://www.imdb.com/es-es/title/tt0983414/?ref_=tt_eps_tp.

Baillie, Jonhathan and Butcher, Ellen. Priceless or Worthless? The World's Most Threatened Species. United Kingdom: Zoological Society of London, 2012.

Borsò, Vittoria. "Bio-poetics and Dynamic Multiplicity of Bios: How Literature Challenges the Politics, Economics and Sciences of Life". In *Life after Literature, Perspectives on Biopoetics in Literature and Theory*, edited by Zoltán Kulcsár-Szabó, Tamás Lénárt, Attila Simon and Roland Végsó, 17–32. Switzerland: Springer, 2020.

Braverman, Irus. "En-listing life. Red is the color of threatened species lists". In *Critical Animal Geographies*, eds. Gillespie K. and Collard R.C. New York: Routledge, 2015.

_____. "The Regulatory Life of Threatened Species Lists". In *Animals, Biopolitics, Law. Lively Legalities*, edited by Irus Braverman, 19–36. New York: Routledge, 2016.

_____. "Anticipating endangerment: the biopolitics of threatened species lists". *BioSocieties* 12, no. 1 (2017): 132–157.

Crutzen, Paul J., and Eugene F. Stoermer. "The 'Anthropocene'." Global Change Newsletter no. 41 (May 2000): 17–18.

Crutzen, Paul J. "The 'Anthropocene'", in Earth System Science in the Anthropocene, edited by Eckart Ehters and Thomas Krafft. Berlin, Heidelberg: Springer, 2006.

Cooke, Brett. "Biopoetics: The New Synthesis". In Biopoetics. Evolutionary Explorations in the Arts, edited by Brett Cooke and Frederick Turner Lexington, 3–25. Kentucky: ICUS, 1999.

Foucault, Michel. *The History of Sexuality: an introduction.* New York: Pantheon Books, 1978.

_____. "Bio-histoire et bio-politique". In Dits et écrits, vol. III, 1976–1979, edited by Daniel Defert y François Ewald, 95–97. París: Gallimard, 1994.

_____. "The Birth of Social Medicine". In Power, Essential Works of Foucault 1954–1984. Edited by James Faubion, 134–156. New York: The New Press, 2000.

_____. Society Must Be Defended. Lectures at the Collège de France, 1975–1976. New York: Picador, 2003.

_____. "The Crisis of Medicine or the Crisis of Antimedicine?". *Foucault Studies* 1, (2004): 5–19. https://doi.org/10.22439/fs.v0i1.562.

_____. Subjectivity and Truth: Lectures at the Collège de France 1980–1981. Basingstoke: Palgrave Macmillan, 2017.

Hacking, Ian. "Genetics, biosocial groups & the future of identity". *Daedalus* 135, no. 4 (2006): 81–95.

Jackson, Rob and Haff, Peter. "A "neoenvironmental" manifesto". *The Ecological Society of America* 9, no. 2, (2011): 87. https://doi.org/10.1890/1540-9295-9.2.87.

Prozorov, Sergei. "From Biopolitics to Biopoetics and Back Again. On a Counterintuitive Continuity in Foucault's Thought". In *Biopolitics and Ancient Thought*, edited by Jussi Backmann and Antonio Cimino, 168–182. Oxford: Oxford University Press, 2022.

Pugliese, Joseph. *Biopolitics of the More-Than-Human. Forensic Ecologies of Violence.* Durham: Duke University Press, 2020.

Rabinow, Paul. "Artificiality and Enlightenment: From Sociobiology to Biosociality". In *Essays on the Anthropology of Reason*, 91–111. New Jersey: Princeton University Press, 1996.

Schenider, Brigit. "Burning worlds of cartography: a critical approach to climate cosmograms of the Anthropocene". *Geo: Geography and Environment* 3, no. 2 (2016), https://doi.org/10.1002/geo2.27.

Weber, Andreas. *Biopoetics. Towards an Existential Ecology.* Dordrecht: Springer, 2016.

Wolfe, Cary. "Foreword". In Extinction Studies: Stories of Time, Death, and Generations, edited by Deborah Bird Rose, Thom van Dooren, and Matthew Chrulew, vii-xvi. New York: Columbia University Press, 2017.

Yelin, Julieta. "Biopoéticas para las biopolíticas. Una introducción". *Uni(+di)versidades* 3, no. 3, (2015): 1–10.

Youatt, Raffi. "Counting species: Biopower and the Global Biodiversity Census", *Environmental Values* 17, (2008): 393–417. https://doi.org/10.3197/096327108X343149.

Chapter 5

Albrecht, Glenn. "Exiting the Anthropocene and Entering the Symbiocene". *Minding Nature* 2, no. 1 (2016): 12–16.

Gilbert, Scott F., Sapp, Jan and Tauber Alfred I. "A Symbiotic View of Life: We Have Never Been Individuals". *The Quarterly Review of Biology* 87, no. 4 (2012) 325–341.

Haraway, Donna. *When species meet.* Minneapolis: University of Minnesota Press, 2008.

_____. *Staying with the trouble: Making kin in the Chthulucene.* Durham: Duke University Press, 2016.

Kozo-Polyansky, Boris Mikhaylovich. *Symbiogenesis: a new principle of evolution.* Edited by Victor Fet and Lynn Margulis. Cambridge: Harvard University Press, 2010.

Margulis, Lynn and Sagan, Dorion. *Acquiring Genomes: A Theory of the Origin of Species.* New York: Basic Books, 2002.

Singh, Vandana. *Ambiguity Machines and Other Stories.* Easthampton: Small Beer Press, 2018.

Stengers, Isabelle. "Introductory Notes on an Ecology of Practices". *Cultural Studies Review* 11, no 1 (2005): 183–196. https://doi.org/10.5130/csr.v11i1.3459.

_____. Cosmopolitics I. Minneapolis: University of Minnesota Press, 2010.

Chapter 6

Blad, Sylvia. "The impact of 'Anthropotechnology' on Human Evolution". *Techné* 14, no. 2, (2010): 72–87.

Díaz, Santiago. "El biopoder de la biotecnología o el biotecnopoder. Aportes para una bio(s)ética." *Ludus Vitalis* 19, no. 36 (2011): 193–211.

Dupré, John. "The Polygenomic Organism". In *Postgenomics, Perspectives on Biology after the Genome*, edited by Sarah S. Richardson and Hallam Stevens, 56–72. Durham and London: Duke University Press, 2015.

Folkers, Andreas and Opitz, Sven. "Low-carbon cows: From microbial metabolism to the symbiotic planet". *Social Studies of Science* 52, no. 3, (2022): 330–352. https://doi.org/10.1177/03063127221077987.

Foucault, Michel. *The History of Sexuality: an introduction.* New York: Pantheon Books, 1978.

_____. "Bio-history and bio-politics". *Foucault Studies* 18, (2014): 128–130.

Gilbert, Scott F., Sapp, Jan, and Tauber, Alfred I. "A Symbiotic View of Life: We Have Never Been Individuals". *Quarterly Review of Biology* 87, no. 4, (2012): 325–341. https://doi.org/10.1086/668166.

Hayles, Katherine N. *How We Became Posthuman: Virtual Bodies in Cybernetics, Literature, and Informatics.* Chicago: Chicago University Press, 1999.

HKW 100 Years of Now. 2015. "Triggers: Introducing the Technosphere | Birgit Schneider." YouTube. October 20, 2015. https://www.youtube.com/watch?v=Iq3CLHnbD0w.

Jones, Clive G., Lawton, John H., and Shachak, Moshe. "Organisms as Ecosystem Engineers". *Oikos* 69 (1994): 373–386. https://doi.org/10.2307/3545850.

Lemke, Thomas. "Beyond Foucault: From Biopolitics to the Government of Life." In Governmentality: Current Issues and Future Challenges, edited by Ulrich Bröckling, Susanne Krasmann, and Thomas Lemke, 165–84. New York: Routledge, 2011.

Lemke, Thomas. *The Government of Things. Foucault and the New Materialisms.* New York: New York University Press, 2021.

Malabou, Catherine. "One Life Only: Biological Resistance, Political Resistance". *Critical Inquiry* 42, (2016): 429–438.

Margulis, Lynn. *Symbiotic Planet: A New Look at Evolution.* New York: Basic Books, 1988.

Margulis, Lynn and Sagan, Dorion. *Acquiring Genomes: A Theory of the Origin of Species.* New York: Basic Books, 2002.

Opitz, Sven and Folkers, Andreas. "Symbiotic engineering: insects, microbes, and the space of vector control". *BioSocieties* 20 (2025), 481–503.

Schilthuizen, Menno. *Darwin Comes to Town: How the Urban Jungle Drives Evolution.* New York: Picador, 2018.

Sloterdijk, Peter. "Rules for the Human Park: A Response to Heidegger's" "Letter on 'Humanism'". In Not saved, Essays after Heidegger, 193–216. Cambridge: Polity Press, 2017.

_____. "The Domestication of Being. The Clarification of the Clearing". In Not saved. Essays after Heidegger, 89–148. Cambridge: Polity Press, 2017.

_____. "Wounded by Machines. Toward the Epochal Significance of the Most Recent Medical Technology". In Not saved. Essays after Heidegger, 217–236. Cambridge: Polity Press, 2017.

Thales' Well. 2025. "On Life, AI and Technosymbiosis with Katherine Hayles." YouTube. June 2, 2025. https://www.youtube.com/watch?v=avFgDTiLTjc.

Wobbes, Theo. "The Concept of Symbiosis Applied to the Human Technological Culture". *Technophany* 3, no. 1, (2025): 1–20. https://doi.org/10.54195/technophany.18586.

Zalasewics, Jan. "The unbearable burden of the technosphere". The *UNESCO Courier* 2, no. 873 (2018): 15–17.

Žukauskaitė, Audronė. "Immunity and Symbiopolitics". *Athena* 17, (2022): 43–56. https://doi.org/10.53631/Athena.2022.17.4.

Chapter 7

Butler, Octavia E. *Parable of the Sower.* London: Headline, 2019.

_____. *Parable of the Talents.* London: Headline, 2019.

Faria, Catia and Paez, Eze. "It's Splitsville: Why Animal Ethics and Environmental Ethics Are Incompatible". American Behavioral Scientist (2019): 1–14. https://doi.org/10.1177/000276421983046.

Noske, Barbara. "Two Movements and Human-Animal Continuity: Positions, Assumptions, Contradictions". *Animal Liberation Philosophy and Policy Journal* 2, no. 1 (2004): 15–26.

Chapter 8

Adams, Max. *Trees of Life.* Oxford: Princeton University Press, 2019.

Chivian, Eric and Bernstein, Aaron. "How is biodiversity threatened by human activity". In *Sustaining Life: How Human Health Depends on Biodiversity*, edited by Eric Chivian and Aaron Bernstein, 28–74. Oxford: Oxford University Press, 2008.

Dneprov, Anatoly. "Crabs on the Island". In The Molecular Café. Science-fiction Stories, 29–58. Moscow: MIR Publishers, 1968.

Haff, Peter. "The technosphere and its relation to the anthropocene". In *The Anthropocene as a Geological Time Unit: A Guide to the Scientific Evidence and Current Debate*, edited by Jan Zalasiewicz, Colin N. Waters, Mark Williams, Colin Summerhayes, 138–143. Cambridge: Cambridge University Press, 2019.

Katz, Eric. "Preserving the distinction between nature and artifact". In *The Ideal of Nature: Debates about Biotechnology and the Environment*, edited by G. E. Kaebnick, 71–82. Baltimore: The Johns Hopkins University Press, 2011.

Lewontin, Richard and Levins, Richard. "Organism and environment". *Capitalism Nature Socialism* 8, no. 2 (1997): 95–98. https://doi.org/10.1080/10455759709358737.

López-Corona, Oliver and Magallanes-Guijón, Gustavo. "It Is Not an Anthropocene; It Is Really the Technocene. Names matter in Decision Making Under Planetary Crisis". *Front. Ecol. Evol.* 8, no. 214 (2020): 1–5. https://doi.org/10.3389/fevo.2020.00214.

Nancy, Jean-Luc. "Note sur le term de 'biopolitique'" In La création du monde ou la mondialisation. 137–143. Paris: Galilée, 2002.

Schenider, Brigit. "Burning worlds of cartography: a critical approach to climate cosmograms of the Anthropocene". *Geo: Geography and Environment* 3, no. 2 (2016), https://doi.org/10.1002/geo2.27.

Sloterdijk, Peter. "Nature to-be-made. The Crucial Subject of Modern Times". In Crítica do Contemporâneo: Política, edited by Rui Mota Cardoso, 195–218. Portugal and Porto: Fundação Serralves, 2007.

Thoreau, Henry David. *Walking.* San Francisco: Harper San Francisco, 1994.

Chapter 9

Adler, Paul (ed). *The Oxford Handbook of Sociology and Organization Studies.* London: Oxford University Press, 2009.

Arellano-Gault, David and Arturo, Del Castillo. *Promises and perils of compliance.* Berlin: De Gruyter, 2023.

Arellano-Gault, David. *Las trampas de la decisión.* Mexico: Fontamara, 2022.

Blake, Jonathan S. and Nils, Gilman. Children of a modest star. Planetary thinking for an age of crisis. Stanford: Stanford University Press, 2024.

Bologh, Roslyn. *Love or greatness. Max Weber and masculine thinking. A feminist inquiry.* New York: Routledge, 1990.

Bromley, Patricia and John, Meyer. *Hyper-organization.* Oxford: Oxford University Press, 2015.

Burrell, Gibson & Gareth, Morgan. *Sociological Paradigms and Organisational Analysis.* New York: Routledge, 1979.

Chakrabarty, Dipesh. *The Climate of History in a Planetary Age.* Chicago: University of Chicago Press, 2021.

Chen, Katherine. *Enabling creative chaos. The organization behind the Burning Man event.* Chicago: Chicago University Press, 2009.

Cohen, Michael, James, March and Johan, Olsen. "People, problems, solutions and the ambiguity of relevance". In Ambiguity and choice in organizations, edited by James March and Johan Olsen, 24–37. Bergen: Universitestforlager, 1976.

Crozier, Michel and Erhard, Friedberg. The actor and the system. Chicago: Chicago University Press, 1989.

Dewey, John. *The public and its problems.* New York: Henry Holt, 1927.

Escobar, Arturo. *Designs for the pluriverse: Radical interdependence, autonomy, and the making of worlds.* Durham, NC: Duke University Press, 2018.

Goldberg, Ken and Mo, Zhou. *DebateCAFE V1.0: Incentivizing Articulation and Consideration of Adversarial Arguments.* Berkeley: University of California Technical Report No. UCB/EECS-2017-197, 2017.

Habermas, Jurgen. *The theory of communicative action.* London: Polity, 1991.

Hardt, Michael and Antonio, Negri. *Assembly.* Oxford: Oxford University Press, 2019.

Kiel, Douglas. *Complexity in Government.* New York: Jossey-Bass, 1994.

Kropotkin, Peter. *Anarchism.* London: Dover Publication, 2002.

March, James. *Explorations in organizations.* Stanford: Stanford University Press, 2008.

_____. *Handbook of Organizations.* New York: Routledge, 2013

Morel, Christian. *Les décisions absurdes.* Paris: Gallimard, 2014.

Ostrom, Elinor. *Governing the commons: The evolution of institutions for collective action.* Cambridge: Cambridge University Press, 1990.

Shister, Neil. *Radical ritual. How Burning Man changed the world.* New York: Counterpoint, 2019.

Stoney C, Winstanley D. "Stakeholding: confusion or utopia? Mapping the conceptual terrain." *Journal of Management Studies* 38, no.5. (2001): 603–626.

Tallberg, Linda and Lindsey, Hamilton. *The Oxford Handbook of Animal Organization Studies.* Oxford: Oxford University Press, 2023.

Tetlock, Philip and Dan, Gardner. *Superforcasting: the Art and Science of Prediction.* New York: Crown, 2016.

Tronto, Joan C. *Caring democracy: Markets, equality, and justice.* New York: New York University Press, 2013.

Tsoukas, Haridimos and Robert, Chia. "On Organizational Becoming: rethinking organizational change". *Organization Science* 13, no. 5 (2002): 567–582. http://dx.doi.org/10.1287/orsc.13.5.567.7810.

Tyler, Melissa. *Judith Butler and Organization Theory.* New York: Routledge, 2020.

Weick, Karl. *Making sense of the organization.* Chichester: Wiley, 2002.

Zamudio-González, Laura and David, Arellano-Gault. *International Organizational Anarchy.* Scham: Palgrave Mac Millan, 2025.

INDEX

www.ingramcontent.com/pod-product-compliance
Ingram Content Group UK Ltd.
Pitfield, Milton Keynes, MK11 3LW, UK
UKHW040232280826
474579UK00002B/6

9 781839 998485